SWEDEN

Sk.	Skåne	Vrm.	Värmland
Bl.	Blekinge	Dlr.	Dalarna
Hall.	Halland	Gstr.	Gästrikland
Sm.	Småland	Hls.	Hälsingland
Öl.	Öland	Med.	Medelpad
Gtl.	Gotland	Hrj.	Härjedalen
G. Sand.	Gotska Sandön	Jmt.	Jämtland
Ög.	Östergötland	Ång.	Ångermanland
Vg.	Västergötland	Vb.	Västerbotten
Boh.	Bohuslän	Nb.	Norrbotten
Dlsl.	Dalsland	Ås. Lpm.	Åsele Lappmark
Nrk.	Närke	Ly. Lpm.	Lycksele Lappmark
Sdm.	Södermanland	P. Lpm.	Pite Lappmark
Upl.	Uppland	Lu. Lpm.	Lule Lappmark
Vstm.	Västmanland	T. Lpm.	Torne Lappmark

NORWAY

Ø	Østfold	HO	Hordaland
AK	Akershus	SF	Sogn og Fjordane
HE	Hedmark	MR	Møre og Romsdal
O	Opland	ST	Sør-Trøndelag
B	Buskerud	NT	Nord-Trøndelag
VE	Vestfold	Ns	southern Nordland
TE	Telemark	Nn	northern Nordland
AA	Aust-Agder	TR	Troms
VA	Vest-Agder	F	Finnmark
R	Rogaland		

n northern s southern ø eastern v western y outer i inner

FINLAND

Al	Alandia	Kb	Karelia borealis
Ab	Regio aboensis	Om	Ostrobottnia media
N	Nylandia	Ok	Ostrobottnia kajanensis
Ka	Karelia australis	ObS	Ostrobottnia borealis, S part
St	Satakunta	ObN	Ostrobottnia borealis, N part
Ta	Tavastia australis	Ks	Kuusamo
Sa	Savonia australis	LkW	Lapponia kemensis, W part
Oa	Ostrobottnia australis	LkE	Lapponia kemensis, E part
Tb	Tavastia borealis	Li	Lapponia inarensis
Sb	Savonia borealis	Le	Lapponia enontekiensis

USSR

Vib Regio Viburgensis Kr Karelia rossica Lr Lapponia rossica

FAUNA ENTOMOLOGICA SCANDINAVICA
Volume 21 1988

Stoneflies (Plecoptera) of Fennoscandia and Denmark

by

A. Lillehammer

E. J. Brill/Scandinavian Science Press Ltd.
Leiden · New York · København · Köln

Library of Congress Cataloging-in-Publication Data

Lillehammer, A.
Stoneflies (Plecoptera) of Fennoscandia and Denmark / by A. Lillehammer.
165 p. 14.5×21 cm. - (Fauna Entomologica Scandinavica. ISSN 0106-8377; v. 21).

Bibliography: p. 156

Includes index

ISBN 90 04 08695 1

1. Stoneflies-Scandinavia-Classification. 2. Stoneflies-Scandinavia.
3. Insects-Classification. 4. Insects-Scandinavia-Classification. I. Title.
II. Series.

QL530.24.S34L55 1988 595.7'35'0948-dc19 88-5075 CIP.

ISBN 90 04 08695 1
ISSN 0106-8377

Printed in Denmark by Vinderup Bogtrykkeri A/S

Contents

Introduction

The Plecoptera is a small order of insects distributed over most of the world. The highest concentration of species is found in the temperate zones of both hemispheres.

It is a well-investigated group of insects and the first world catalogue of species was published by Illies (1966). Zwick (1973) made certain revisions and added more phylogenetic information.

According to Zwick (1973) there are more than 1700 species, placed in two suborders: the Antarctoperlaria, restricted entirely to the southern hemisphere, and the Arctoperlaria, found in both the northern and the southern hemispheres.

Our knowledge of plecopteran species is rapidly increasing, especially of their autecology. All the developmental stages of several species have been studied, i.e. eggs, nymphs and adults. These studies also include the mating behaviour, during which both sexes produce drumming signals. The information derived from these studies has been used to solve problems within the fields of systematics and zoogeography.

The European distribution has been discussed by Illies (1953, 1967, 1978). He divides the European landscape into 25 geographical regions, Fennoscandia and Denmark representing 5 of these. The European distribution has also been discussed by Raušer (1962, 1971) and by Zwick (1981, 1984), while Aubert (1959) and Hynes (1941) provide information about Switzerland and Great Britain respectively. Brekke (1941), Brinck (1949), Meinander (1965, 1975, 1980, 1984), Jensen (1951), Hynes (1953), Kaiser (1972), Lillehammer (1974b, 1985a), and Tobias (1973, 1974) have published studies on the distribution of the plecopteran species in Fennoscandia or Denmark.

Both adults and nymphs are classified on external morphological characters. The nymphs differ only slightly from the adults, the chief differences being the absence of wings and genitalia. Brinck (1949) has published keys to and much information on the species occurring in Sweden, and Esben-Petersen (1910b) has done the same for the Danish fauna. However, none of the publications mentioned above contain information about the nymphs and adults of all of our species, and many of the identification characters used in the keys have been found to vary so widely that they are of only limited use. The time is therefore ripe for a modern revision.

Diagnostic characters

The suborder Arctoperlaria is divided into two groups, the Systellognatha and the Euholognatha. The Systellognatha is characterised by the following synapomorphic characters: adults possess reduced mandibles; the egg chorium is thick and sclerotised and possesses both an anchor plate and an egg collar (Figs 1, 2); the male has a ventral opening on the 10th tergum that bears the retractable epiproct; a sclerotised bridge ex-

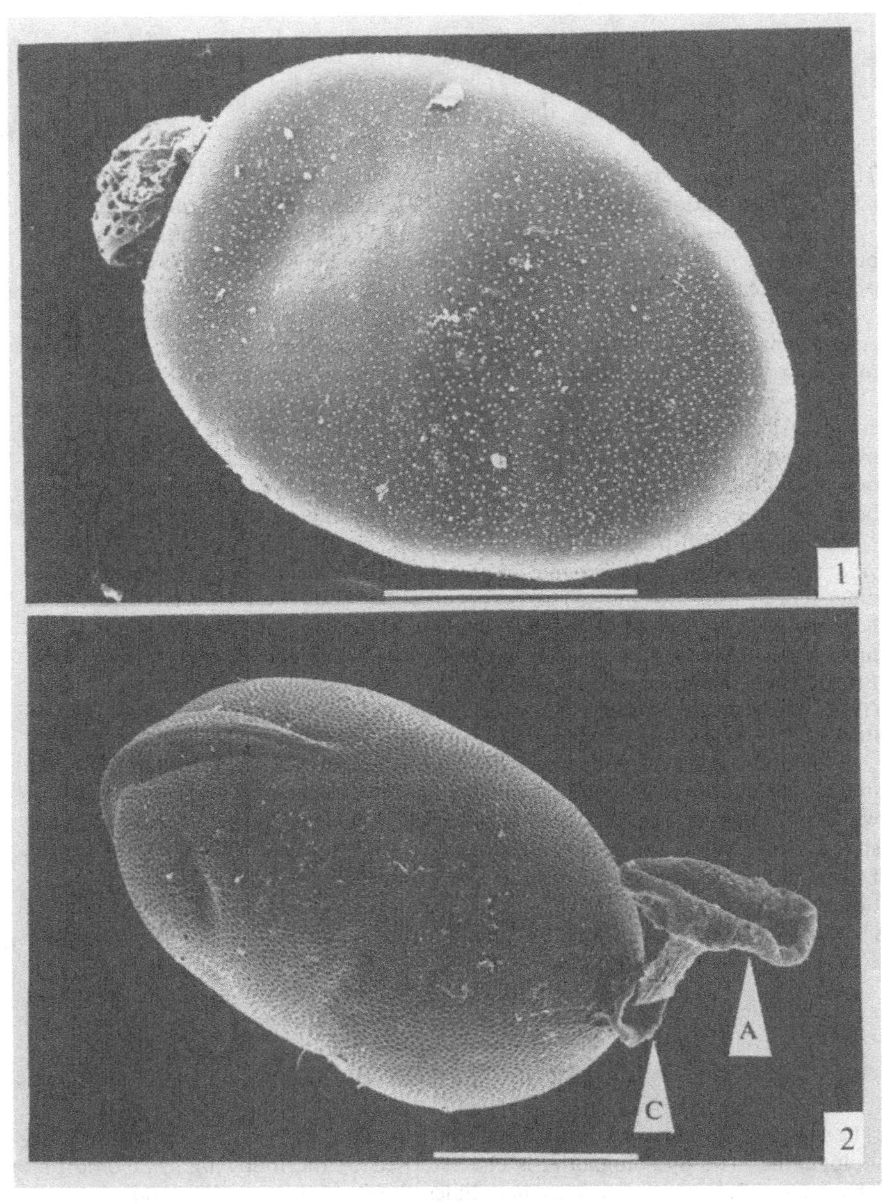

Figs 1, 2. SEM photos of eggs of 1: *Diura nanseni* (Kempny) and 2: *Isoperla obscura* (Zett.). A = anchor plate, C = egg collar. Scale: 100 μm.

ists between the epimer and the nota of the pterothorax. The following characters characterise the Euholognatha: adults possess well developed and functionable mandibles; the egg chorium is thin and membranous, and not provided with an anchor plate and a collar; the Corpus allatum is unpaired.

Eggs

The egg chorium of Systellognatha species may be darkly pigmented, as in *Dinocras cephalotes*, *Arcynopteryx compacta* and the *Diura* species, or only weakly pigmented, as in *Isoperla obscura* and *Siphonoperla burmeisteri*. Characteristic microstructures, including the micropyles, are often visible in SEM photos of the chorium (Figs 1, 2). Eggs of Euholognatha species have an unpigmented, membranous chorium, that makes the interpretation of SEM studies difficult. Egg structure, as studied by SEM photography, has been used to ascertain phylogenetic relationships, as well as to support specific differences in the Nearctic Pteronarcynidae (Stark & Szczytko, 1982). This method has not yet been used to any great extent in studies of European Plecoptera, but Stark *et al.* (1986) have given some information about Western Palearctic species, and Lillehammer & Økland (1987) have used the egg chorium structures in a key to the Fennoscandian *Isoperla* species.

Nymphs

Stoneflies develop into adults through several nymphal instars and some specific differences exist: *Pteronarcys proteus* has 12 instars (Holdsworth, 1941) and *Dinocras cephalotes* 33 (Schoenemund, 1912). Wu (1923) mentions the presence of 22 instars in *Nemoura* species, but Elliott (1984) recorded twelve to fourteen instars for *Nemoura avicularis*, and Harper (1973) sixteen for *Nemoura trispinosa* Claassen.

The eggs of 25 species of the Fennoscandian fauna have so far been successfully hatched in the laboratory to first instar nymphs. Although these nymphs belong to known species, it is at present difficult to arrange them in keys based on morphological characters. One species with an easily identifiable first instar nymph is *Dinocras cephalotes*, which has gills situated at the apex of the abdomen (Fig. 52). The body size of the first instar nymph varies according to species. The smallest are about 0.5 mm long and the largest about 1.7 mm. Until the nymphs have attained a body length of about 3 mm, identification to species level is difficult.

The first instar nymphs of all the studied species have nine-segmented antennae, and all the Systellognatha species also possess three-segmented cerci. This character they share with species of the families Taeniopterygidae, Nemouridae and Capniidae, while the first instar nymph of Leuctridae species have four-segmented cerci. Information about the antennae and cerci of first instar nymphs has also been given by Septon & Hynes (1982) and Harper (1973).

The head (Fig. 3) of all nymphal instars is large, flattened dorso-ventrally, and with large laterally placed eyes. Ocelli are most often present, but may be absent, as in several *Nemoura* species.

The mouth parts are biting, and the shape of the following elements is often used for identification purposes: the mandible, the maxilla with the lacinia, the galea, and

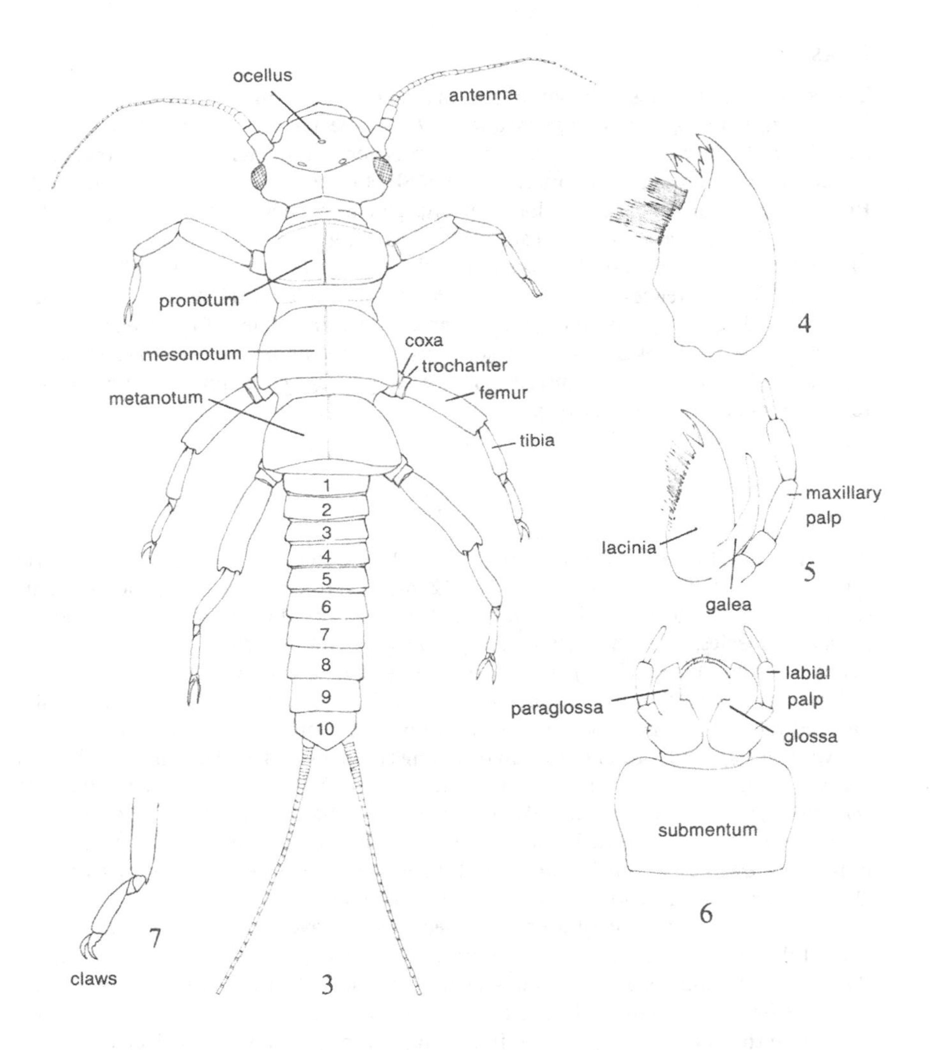

Figs 3-7. Nymph of *Isoperla obscura* (Zett.). - 3: habitus, dorsal view; 4: mandible; 5: maxilla; 6: labium and submentum; 7: tarsus.

the maxillary palp, and finally the labium with the glossa, the paraglossa, the submentum, and the labial palp. The shape of the tentorial callosity is also sometimes used. The pattern of the pigmentation of the dorsal side of the head is important for the separation of *Isoperla* species.

The thorax is covered dorsally by three plates, the pronotum, the mesonotum, and the metanotum, laterally by the pleura, and ventrally by the sterna. The legs have a lateral position and are composed of the coxa, the trochanter, the femur, the tibia, and the tarsus with three segments, terminating in two claws (Fig. 7).

The abdomen has ten segments, terminating in a pair of multi-segmented cerci. Two plates, the paraprocts, are present on the ventral side. Each abdominal segment consists basically of a tergum (dorsally) and a sternum (ventrally). On the first two abdominal segments the tergum and the sternum are always separated, whereas in segments 3-9 they may or may not be entirely fused; on the tenth segment they are always fused.

Hairs and bristles are present on the body, and the arrangement of these are important for the identification. The bristles on the pronotum, on the legs, on the dorsal side of the abdomen (terga), and on the segmented cerci are most commonly used for identification purposes.

The nymphs of several species also have external gills, which may be filiform or sausage-shaped. In Fennoscandian species gills may occur laterally on the coxa, as in *Taeniopteryx nebulosa,* on the apex of the abdomen, as in *Dinocras cephalotes,* or on the prosternum, as in the *Amphinemura* and *Protonemura* species.

Adults

The adult body is usually flattened dorso-ventrally. The head is large and bears biting mouth-parts, which are well-developed in the Euholognatha but reduced in the Systellognatha. The antennae are long and filiform. The eyes are large and well-developed; usually three ocelli are present. The mandibles are reduced in most of the Systellognatha, but well-developed in the Euholognatha. The maxilla is divided into two parts, an outer galea and an inner lacinia. The maxillary palps are 5-segmented; they are filiform in the Euholognatha and bristle-like in the Systellognatha. The form of labium varies. In the Systellognatha the paraglossa is large and the glossa reduced, whereas in the Euholognatha they are of about equal size. The labial palps have 3 segments; they are short and blunt in the Euholognatha, long and filiform in the Systellognatha (Fig. 3). The prothorax bears a sclerotised pronotum, and the meso- and metathorax each comprise a pair of wings.

The fore- and the hindwings of stoneflies are very different in shape (Fig. 8). The hindwings usually have a well-developed anal area. This may be much reduced in some species and of almost the same size as the forewing, as in *Siphonoperla burmeisteri* (Fig. 9). The males of some species have short wings. Among the Fennoscandian Plecoptera these include *Diura bicaudata* (Fig. 12), *Arcynopteryx compacta, Perlodes dispar, Isoperla difformis, Dinocras cephalotes* (Fig. 11), *Capnia vidua* and *C. bifrons.*

In some species the wings may be more or less reduced. Sometimes only some members of a population have such short wings, sometimes the whole population. Short-wingedness seems to be both genetically determined and environmentally induced (Lillehammer, 1985d). Variation in wing length is often accompanied by a variation in the characters of the genitalia (Lillehammer, 1974a). The reduction in wing length is first noticeable at the apex and is often followed by irregularities in the wing ribs.

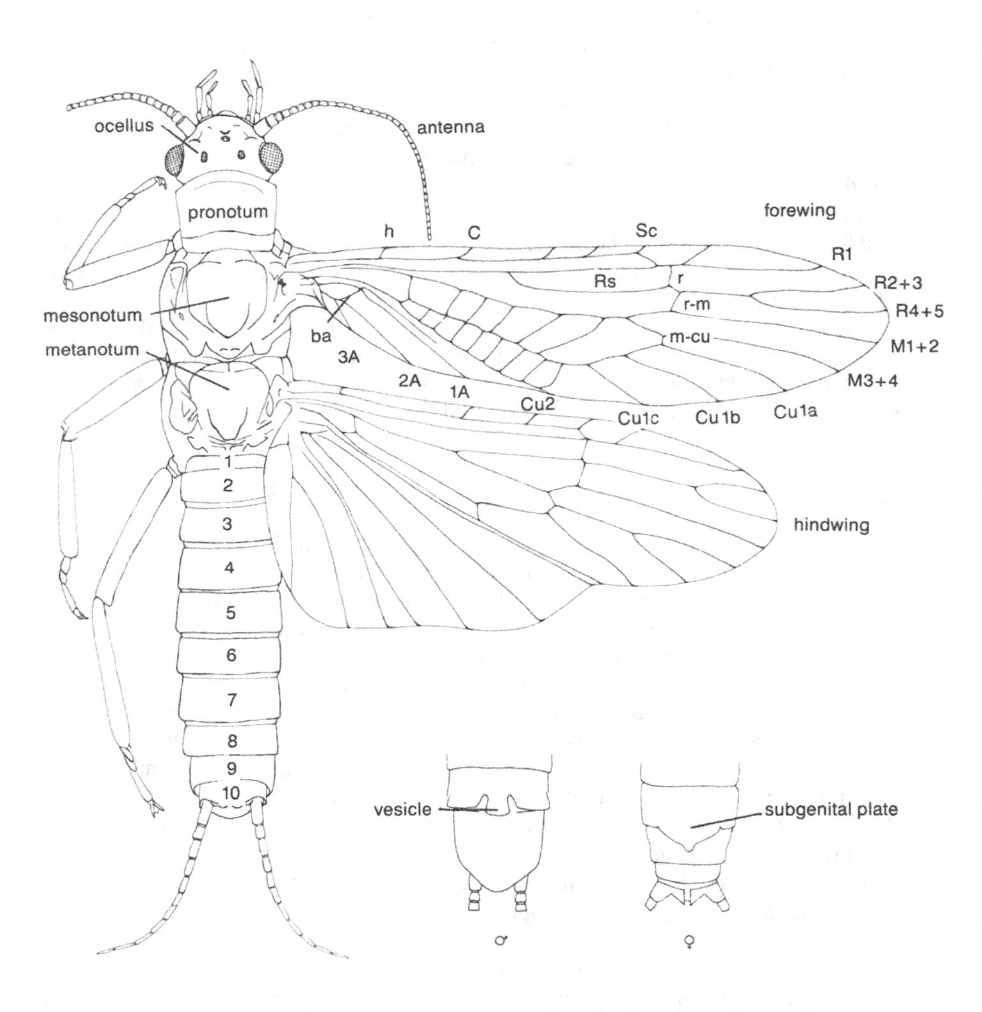

Fig. 8. Adult of *Isoperla obscura* (Zett.). For abbreviations for the wing veins, see the text.

The body length of our stonefly species varies widely. The largest is *Dinocras cephalotes,* the females of which can attain a body length of 30 mm. The smallest is *Capnopsis schilleri* with a female body length of only 4-5 mm. However, a wide intraspecific variation in body length is often found and some species regularly include populations that are dwarfs such as some of the high altitude populations of *Amphinemura standfussi.* Such a reduction in body length is often accompanied by a reduction in the size of all the other parts of the body, such as head width, head length, femur length etc., including also wing length (Lillehammer, 1985d).

The form of the wing and of the wing ribs are taxonomically important. The costa (C) forms the anterior margin of the wings (Fig. 8), while the subcosta (Sc) is bifurcated and fused with the radius (R1). The radius (R) has two branches, a front branch (R1) and a hind branch (R2), the latter being bifurcated (R2+3 and R4+5) and may even be secondarily subdivided. The media (M) is also bifurcated (M1+2 and M3+4). Also

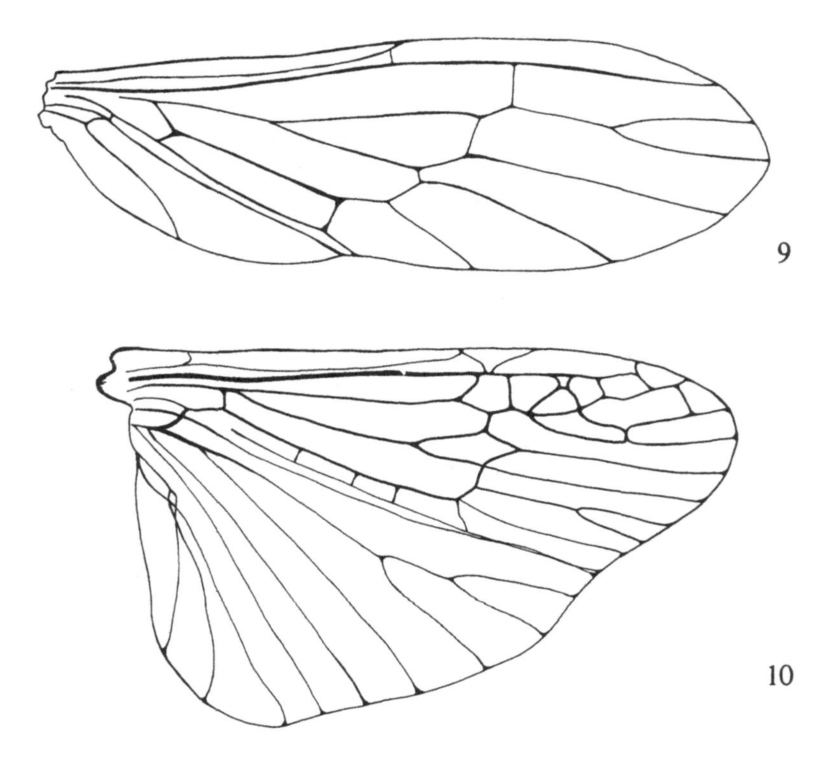

Figs 9, 10. Hindwing of 9: *Siphonoperla burmeisteri* (Pict.) and 10: *Arcynopteryx compacta* (McL.), the former with reduced anal area.

the cubitus (Cu) is bifurcated, Cul bending forwards and often fused with M3+4. Cu often bears accessory veins, which are termed Cula, Culb, etc. The anal veins (1A, 2A, 3A) are often branched. Cross-veins also occur, viz., (r) the radial cross-vein, (h) the humeral cross-vein, (r-m) the radio-medial cross-vein, (m-cu) the medio-cubital cross-vein, and (ba) the anal cross-vein between 1A and 2A.

The abdomen usually has ten segments. In the Systellognatha the 1st is often fused to metasternum, and in the Euholognatha the 10th is often reduced. The genitalia of both sexes are very diverse in form. In the females the sternum of the 7th, 8th or 9th abdominal segment is modified to form a subgenital plate. The oviduct is short and extends from the posterior part of the 7th or 8th sternum (Brinck 1956). In the Nemouridae the opening is situated on the posterior margin of the 7th segment, which protrudes posteriorly to form a subgenital plate (Fig. 179). In some species there may be sclerotised structures below the plate, as in some *Nemoura* species (Fig. 182). In other stoneflies, such as *Leuctra* species, the opening is situated behind the 8th sternum, and the subgenital plate is laterally lobed (Figs 249-252). *Leuctra* spp. females may also

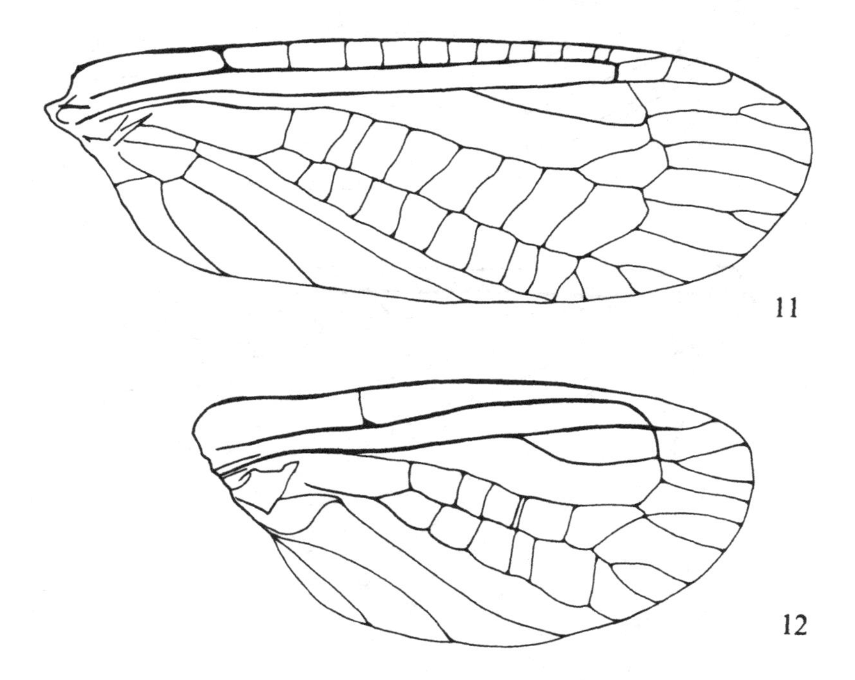

Figs 11, 12. Forewing of short-winged males of 11: *Dinocras cephalotes* (Curt.) and 12: *Diura bicaudata* (L.), the former with numerous cross-veins between costa and subcosta.

possess a sclerotised spermatheca (Lillehammer, 1974a: fig. 48). In the Perlodidae and Chloroperlidae the anterior part of the 8th sternum is modified into a subgenital plate of variable size (Figs 62-67). Sternum 9 is usually unmodified, but in Taeniopterygidae species it is modified to form a postgenital plate (Figs 124-127).

As in the females the 1st sternum of Systellognatha males is often fused to the metasternum, and the 10th sternum is often reduced in the Euholognatha. The male tergal processes are important for the identification of Leuctridae species. The processes start from the 6th tergum backwards (Figs 245-248). In the family Capniidae knobs and other structures may occur on terga 6-8 (Figs 223-227). Such structures are usually not present in species of Systellognatha. The 9th segment carries the male gonophore and the sternum is often produced into a subgenital plate, which covers the 10th sternum. In the Taeniopterygidae (Figs 120-123) the plate is upcurved. The 9th tergum is usually unmodified in those species that do not possess a well-developed supra-anal lobe, such as many species of Systellognatha. In other species, such as of the family Leuctridae, well-developed structures may be present (Figs 245-248). The males of Taeniopterygidae, Nemouridae and Leuctridae possess a ventral lobe on the 9th sternum. In the Capniidae, this organ may be present in some species, such as *Capnia bifrons,* and absent in others, such as *C. atra.* A ventral lobe is well developed in *Isoperla,* but occurs on the 8th sternum (Figs 68-70). It is reduced or absent in the other species of Perlodidae. The ventral lobe of *Leuctra* and *Nemoura* species has been described by Ruprecht (1976), but the function of this organ is still debatable. The 10th segment is very diverse in form. In some genera it may consist of a complete ring, but usually the sternum is more or less reduced and the tergum modified in different ways. As in the females, two paraprocts are situated behind the 10th sternum. There is also a dorsal structure, the epiproct, situated behind the 10th tergum, and which often has a complex structure and projects upwards, as in Nemouridae species (Figs 208-210, 156-171). The epiproct may have well sclerotised internal structures, such as shown in figs 172-178. The epiproct is also strongly sclerotised and well-developed in Capniidae species (Figs 223-228). In species of Chloroperlidae the epiproct bears a characteristic tooth (Figs 104-106). The supra-anal lobe may be well-developed, as in *Leuctra* (Figs 245-248), or reduced such as in *Xanthoperla apicalis.* A complete description of the plecopteran reproductive system has been given by Brinck (1954, 1956, 1970). A wide intraspecific variation in the external characters of the genitalia is found in some species, as mentioned by Kuhtreiber (1934), Brinck (1949), Lillehammer (1974a, 1976). Such differences may also be exhibited by the micro-structures (Lillehammer, 1986a).

The form of the cerci varies greatly in stoneflies. Both sexes of all species of the Systellognatha have long and multi-segmented cerci. Both sexes of species belonging to the families Nemouridae and Leuctridae have small and one-segmented cerci. In the family Capniidae both sexes have multi-segmented cerci, although the number of segments is greatly reduced in *Capnopsis schilleri.* In the family Taeniopterygidae, the males of *Taeniopteryx* and *Brachyptera* have one-segmented cerci, while both sexes of *Rhabdiopteryx* have 4- og 5-segmented cerci. The cerci of the females of *Taeniopteryx* have 6 to 8 segments, and those of *Brachyptera* one segment.

Ecology

Stoneflies in general inhabit running water, but some species are also well adapted to the living in lakes. The local composition of a stonefly fauna will depend in particular on the ability of each species to cope with the prevailing environmental factors during the different stages of its life, from egg to adult.

Eggs

The eggs of all the Fennoscandian stonefly species are normally deposited in water. One exception has been noted in laboratory experiments: *Nemoura viki* sometimes deposited its eggs in water and sometimes on damp moss polsters, when both habitats were available.

The egg incubation period is defined as the time taken by the eggs until 50% of them have hatched into nymphs. The egg incubation period of stoneflies has been studied by several authors, such as Hynes (1941), Brinck (1949), Harper (1973), Lillehammer (1975b, 1985b, 1986 d, 1987a,b), Brittain (1977, 1978), Brittain *et al.* (1984), Haaland

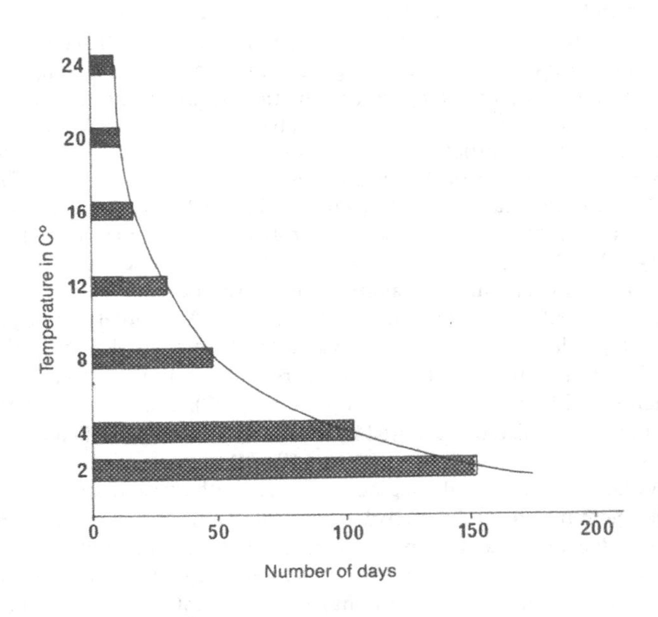

Fig. 13. Correlation between the length of the incubation period of eggs of *Leuctra hippopus* Kempny and the water temperature.

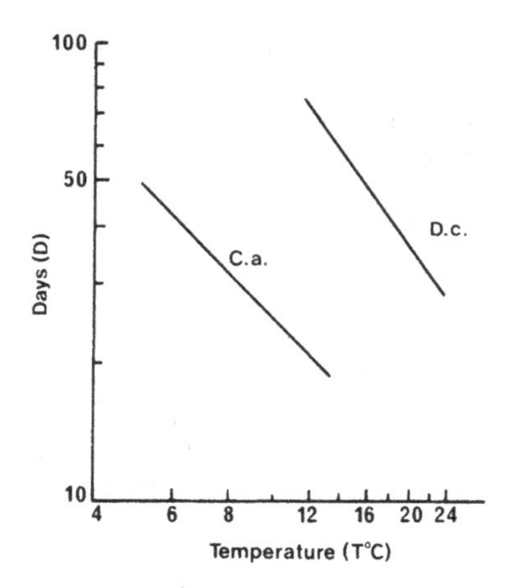

Fig. 14. Correlation between the length of the egg incubation period in two stonefly species and the water temperature, plotted onto a logarithmic scale. C.a. = *Capnia atra* Morton, D.c. = *Dinocras cephalotes* (Curt.).

(1981), Rekstad (1979), Saltveit (1977), Saltveit & Lillehammer (1984), and Zwick (1981). Studies made by Lillehammer *et al.* (1988) have shown that three types of egg incubation occur among the Fennoscandian fauna. 1. Ovovivipary: only in *Capnia bifrons.* 2. Diapause: eggs are deposited in summer and go into a diapause during the autumn. The incubation period lasts for 8-10 months at low temperatures during the winter and hatching takes place in springtime the following year (Lillehammer, 1976, 1985c; Saltveit, 1977; Saltveit & Lillehammer, 1984). *Isoperla obscura, Diura bicaudata, D. nanseni* and *Arcynopteryx compacta* have eggs of this type. Diapausal eggs may react in different ways to the environmental factors, such as water temperature. While the eggs of *Arcynopteryx compacta* can be taken out of diapause in early winter, by first freezing them and then increasing the water temperature, those of *Diura bicaudata* seem to be dependent on other factors, such as the photoperiod (Lillehammer, 1987b). 3. Non-diapause: this is the most common type found among the Fennoscandian stoneflies. The duration of the incubation period is usually short at high water temperatures, and increase with decreasing water temperatures. Non-diapausal eggs usually have an incubation period during the late spring, summer or early autumn, at relatively high water temperatures. The length of the incubation period is greatly dependent upon the water temperature (Fig. 13). This is the most common type of incubation found among the Fennoscandian stonefly fauna, Brittain (1977, 1978, 1983a), Brittain *et al.* (1984), Lillehammer (1975b, 1985a, b, 1986b, 1987b), Saltveit (1977), Saltveit & Lillehammer (1984), Rekstad (1979), Haaland (1981).

For species with non-diapausal eggs, a correlation exists between the egg incubation period (y days) and the water temperature (T°C). This is significantly linear on a logarithmic scale, and can be expressed by the equation: $\log Y = \log a - b \times \log T$, or $Y = aT^{-b}$ (Fig. 14), where a and b are constants (Brittain, 1977). Marked specific differences exist in the degree of temperature dependence. Often these differences are greater between species belonging to the same genus than between the species of different genera. In the genus *Isoperla, I. obscura* has diapausal eggs, while *I. grammatica* and

I. difformis have temperature-dependent eggs, which react significantly differently to the same water-temperatures (Saltveit & Lillehammer, 1984).

The amount of warmth required for egg development can be expressed as day-degrees C. Those of large species, such as *Dinocras cephalotes,* requires as much as 784 day°C to complete development, while those of a smaller species, such as *Nemoura arctica* (Lillehammer, 1986b, 1987c) require only 250 day°C.

Two main types of reaction to the day-degree requirement produced by a change in the water temperature are seen in stonefly eggs. The first type exhibits a negative reaction to a lowering of the temperature. The development time is greatly prolonged, and the day°C requirement at low temperatures is much higher than that at high temperatures. This type is found in a number of species, such as *Leuctra hippopus, Dinocras cephalotes, Siphonoperla burmeisteri* and *Isoperla difformis.* The second type, found in species such as *Amphinemura sulcicollis, Leuctra fusca* and *Leuctra digitata,* is a positive reaction to a lowering of the temperature, whereby the day°C requirement decreases.

In some other species, such as *Nemoura arctica,* the day°C requirement remains at about the same level, despite changes in water temperature.

Thus the same temperature fluctuation may influence the egg development of different species in different ways. This seems to be important as a means of achieving a separation in time of the development of the eggs of closely related species, so that they can co-exist in the same biotope and avoid competition.

The eggs of different non-diapausal species require different water temperatures to initiate their development. The eggs of *Dinocras cephalotes,* for example, require a temperature of 10-12°C before development starts, and have the best chance of hatching success at 16-20°C (Lillehammer, 1986b). The eggs of some other species, such as *Leuctra digitata,* on the other hand, can still develop at a temperature of 2°C, and have the best chance of hatching success at temperatures between 2° and 16°C (Lillehammer, 1985b). When the temperature is 8°C or less, the eggs of *Dinocras cephalotes* may have a period of quiescence that can last for a whole year. This quiescence period can be broken at any time of the year by a rise in the water temperature. The duration of the egg incubation period of this species at 20°C, however, is only 30-35 days (Lillehammer, 1987b). A similar, but obverse, type of quiescence period has been recorded for the eggs of *Amphinemura standfussi* by Saltveit (1977). The eggs developed continuously at temperatures between 4° and 12°C, but went into quiescence when transferred to water temperatures of 16° and 20°C. When held at these higher temperatures for 5 months, they nevertheless continued to develop normally when returned to water at the lower temperatures.

Three types of temperature dependence can be listed: A. Low temperature species achieve full hatching success only at temperatures between 2° and 12° or 16°C. Species such as *Amphinemura standfussi, Leuctra fusca, L. digitata* and *L. nigra* belong to this group (Saltveit, 1977; Haaland, 1981; Lillehammer, 1985b). B. Warm stenothermic species need temperatures of 8° or 12°C, or above for complete success in egg development. Species such as *Isoperla grammatica* (Saltveit & Lillehammer, 1984) and *Dinocras cephalotes* (Lillehammer, 1987c) belong to this type. C. Some stonefly species are eurytherm, achieving complete hatching success at temperatures between 2° and 24°C. *Nemoura cinerea* and *Amphinemura sulcicollis* (Saltveit, 1977; Rekstad, 1979) belong to this type.

Nymphs

Stonefly nymphs occur in both lentic and lotic habitats, in lowland areas as well as at high altitudes, but they are more commonly found in running waters than in lakes. In lakes the highest number of species have been recorded in the northernmost part of Fennoscandia and at high altitudes (Lillehammer, 1985a; Brinck & Froelich, 1960; Brinck & Wingstrand, 1949, 1951), and at high altitudes by Bagge (1965), Bagge & Salmela (1978), Ulfstrand (1975), Ulfstrand *et al.* (1971). Two species are most commonly recorded in stagnant water: *Nemoura viki* in northern and *N. dubitans* in southern Fennoscandia.

The nymphs of stoneflies are usually supposed to require fast-flowing water, rich in oxygen. Recent laboratory studies, (Lillehammer, 1985b, 1986b, 1987c; Brittain, 1977, 1978), however, indicate that this may simply be a preference rather than an absolute requirement. The type of substrate of the habitat seems nonetheless to be very important.

Madsen (1968, 1969) has studied *Brachyptera risi* and *Nemoura flexuosa.* He concluded that the nymphs of the former showed a strong preference for highly-oxygenated running waters, whereas those of *N. flexuosa* could survive under much poorer oxygenation conditions. Those species that live among the detritus on the river bottom substrate are, however, not exposed to running water conditions.

The rate of growth of stonefly nymphs mainly depends on the ambient temperature (Hynes, 1941, 1970; Brinck, 1949) and on the quality and quantity of the food supply (Lillehammer, 1975b; Baekken 1981). Studies of temperature dependence have been made by Harper (1973), Svensson (1966), Bengtsson (1972, 1979), Benedetto (1973b), Iversen (1978), Lillehammer (1975b, 1985b, 1986b), and Brittain *et al.* (1986). Ward & Stanford (1982) have published a review of studies so far made on this topic. In general, nymphal growth is rapid at high and slow at low water temperatures (Fig. 15). That of some species, however, seems to be little influenced by changes in temperature, e.g. *Nemurella pictetii* (Elliott, 1984) and *Capnia atra* (Brittain *et al.,* 1986). Differences in the growth-rates of the different local populations of some species have been recorded (Lillehammer, 1987a; Brittain *et al.,* 1986). In the main, plecopteran nymphs grow continuously during the summer. Some exceptions to this rule occur, however. Khoo (1968b) found that young nymphs of *Capnia bifrons* went into a diapause during the warmest period of the summer in June, when the water temperature was 9.5-13.5°C. The diapause was broken in September. He considered that both the temperature and the photoperiod had an influence on the duration of the diapause.

The food supply as a factor in nymphal growth has been mentioned by several authors. Both Hynes (1941) and Brinck (1949) have analysed the feeding habits of stonefly nymphs. Laboratory studies on the food of stonefly nymphs have been carried out by authors such as Brinck (1949), Lillehammer (1975b, 1976, 1985b), Brittain

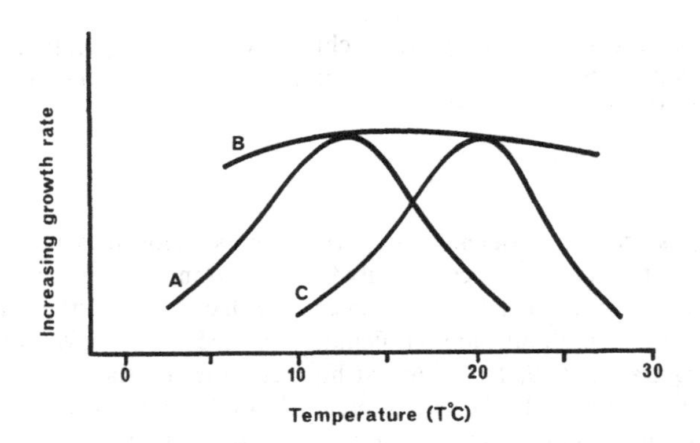

Fig. 15. Different types of growth rate for stonefly nymphs. A: *Leuctra hippopus* Kempny; B: *Capnia atra* Morton; C: a hypothetical type.

(1974, 1978), Saltveit (1977), Rekstad (1979), and Malmquist & Sjøstrøm (1980). Eglishaw (1964) showed that a significant correlation exists between the distribution of some herbivorous stonefly species and the amount of plant detritus present in streams, and both Lillehammer (1974b) and Lillehammer & Brittain (1978) found a correlation, at high altitudes, between the species zonation and the occurrence of willow (*Salix* spp.) in the vegetation. According to Andèrsson & Cummins (1979), both food quality and quantity have an influence on the course of the life cycle of certain stonefly species. Lillehammer (1975b, 1976) has demonstrated that shortage of food led to a prolongation of the life cycles of three stonefly species. The co-existence of related species has been studied by several authors such as Fuller & Stewart (1977, 1979), Sheldon (1980a, b), and Lillehammer (1985c). A changeover from algal and detrital feeding to predation has been studied by Sigfrid & Knight (1976), Fuller & Stewart (1977), Winterbourn (1974), and Lillehammer (1985c). The energy content of the different prey animals taken by nymphs of *Dinocras cephalotes* has been studied by Malmquist & Sjøstrøm (1980).

Hynes (1970) has put forward the hypothesis that microorganisms play a major role in the trophic relationships of stoneflies living in running water. The studies of *Amphinemura sulcicollis* made by Madsen (1974) have verified this hypothesis. Cummins & Klug (1979), in discussing the feeding of aquatic invertebrates, stated that it is difficult to separate the individual effect of water temperature and food quality on nymphal growth.

The results of laboratory studies on water temperature and food supply have shown that both these factors influence the nymphal growth (Lillehammer, 1975b). A lowering of the temperature, or a shortage of food increased the development time, while high temperatures led to a decrease.

At the optimal water temperature, the nymphs of several species grow rapidly until the maturation begins, whereupon growth slows down during the maturation process and until emergence, as shown by *Nemoura arctica.* At low water temperatures growth is slow and gradual during the whole developmental process (Lillehammer, 1986b).

The growth rate of different species, calculated for the premature period, can be expressed by the simple linear regression $G = a + bD$, where a and b are constants, growth (G) in mm and D (days).

The type of substrate has a marked influence on the distribution of stoneflies. The highest species diversity is found on stony bottoms in running water habitats, while a sandy substrate is only occupied by the nymphs of a few species, such as *Siphonoperla burmeisteri, Capnopsis schilleri, Capnia pygmaea, Nemoura avicularis* and *Isoptena serricornis.*

The oxygen consumption of the nymphs of several stonefly species has been studied by Knight & Gaufin (1966) and by Nelson & Garth (1984). They conclude that the nymphs of the large-sized species consume less oxygen per mg. body-weight than those of small species. Differences between species of about equal size have also been recorded (Madsen, 1968, 1969).

The hardness of the water is also thought to be important for some species. The view has been expressed (Hynes, 1977) that the absence of *Capnia bifrons* from softwater streams is due to the lack of calcium ions, or of those of some other essential elements. However, *Capnia bifrons* does occur in several softwater streams in Norway, such as at high altitudes in Øvre Heimdalen (Lillehammer & Brittain, 1978).

Adults

Emergence takes place after the mature nymph has crawled onto the shore. When ready for emergence the exuvia split open along the dorsal side. The thoracic segments are drawn out first, then the head, and finally the abdomen. The wings are held upright until they are dry. The newly-emerged adult is at first light-coloured, gradually becoming darker as it dries.

Once dry, the adults then often fly away to the habitat site they will occupy throughout the rest of their life. Some species move into the cover of vegetation, such as tall herbs, grassland, coniferous or deciduous woodland, where they usually remain until the females are ready for depositing their eggs. Other species hide beneath large stones on the river bank, or in the surrounding area. The adults of most species congregate for feeding, resting and copulation. In any given area, the stonefly populations are more mixed than dispersed (Johnson, 1966). The migration behaviour of *Capnia atra* has been studied by Thomas (1966), Müller (1978), and Müller & Mendl (1980). After emergence, the adults of this species walk across the river bank, still snow-covered, towards the nearest wooded area. They often stay in the shelter of the trees for days or even weeks. There copulation, feeding, and egg maturation take place. The females then fly back to the riverside, and deposit their eggs. Daan & Gustavsson (1973) observed that, above the tree-line, adult *Capnia atra* walked to the nearest point in the environment that offered dense shade, often underneath stones, where they aggregated.

The longevity of the adults may be from only a few days up to a couple of weeks, with marked interspecific differences. Those that have a short life-span manage only to mate and deposit the eggs, without ever feeding. Other species live for a longer time and do feed. The adult females of species with herbivorous nymphs usually feed during this stage of life, on algae, twigs and leaves. How much of this food, or which part of the food they are able to utilise, are still open questions. In marginal habitats where food is scarce the adult females may eat dead insects, as has been observed by Saltveit (1977).

Prior to mating both sexes show special patterns of behaviour, which include making drumming signals (Ruprecht, 1969, 1976; Ruprecht & Gnatzy, 1974) (Fig. 16). These signals are species specific, the local populations even having dialects (Ruprecht, 1972, 1982). The recorded drumming signals have been used as an aid to species identification, as mentioned by Stewart *et al.* (1983). Mating occurs as soon as the two sexes have made contact. The male will copulate several times, while the female of at least some species avoid further contact after copulation.

The flight period of a local population of a stonefly species may last for a much longer period than the life span of the single specimen. There are always some differences in the development rates of the individuals in a population, resulting in an extended emergence period.

The flight period of the same species starts much earlier in the lowland and in the southern parts of Fennoscandia than at high altitudes and in the northern parts.

Brinck (1949) states the flight period for 32 stonefly species from different parts of Sweden. For *Isoperla grammatica,* the first day on which emergent adults were recorded was May 15th in southern Sweden and July 7th in the north.

In the coastal areas of south-western Norway, Lillehammer (1975a) recorded the start of the flight period of *Leuctra hippopus* as March 20th, while at high altitudes (above the tree line) in the inner fjord area the flight period only started on June 10th in the same year. This difference in the timing of the flight period seems mainly to be influenced by differences in the temperature, since the length of the photoperiod was exactly the same in both localities.

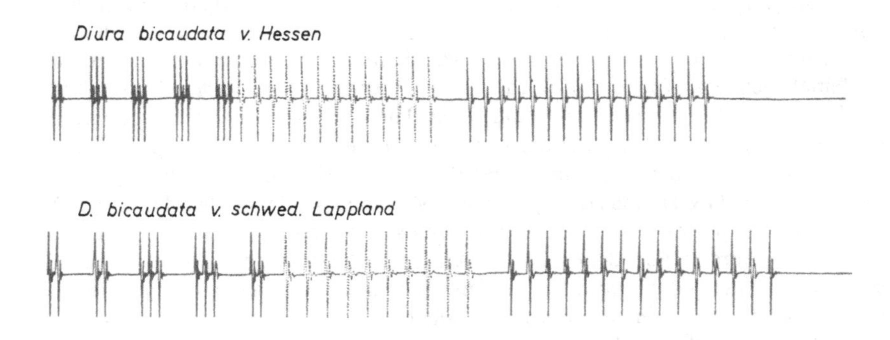

Fig. 16. Oscillogram of drumming signals of females from different populations of *Diura bicaudata* (L.). (After Ruprecht, 1972).

Types of life cycle

The predominant type of life cycle among the Fennoscandian stonefly species is the univoltine one. The majority have a short egg-incubation period during the summer and a long nymphal growth period during the autumn and winter and also extending into the spring of the following year (Fig. 17). Some species, however, have a long egg-

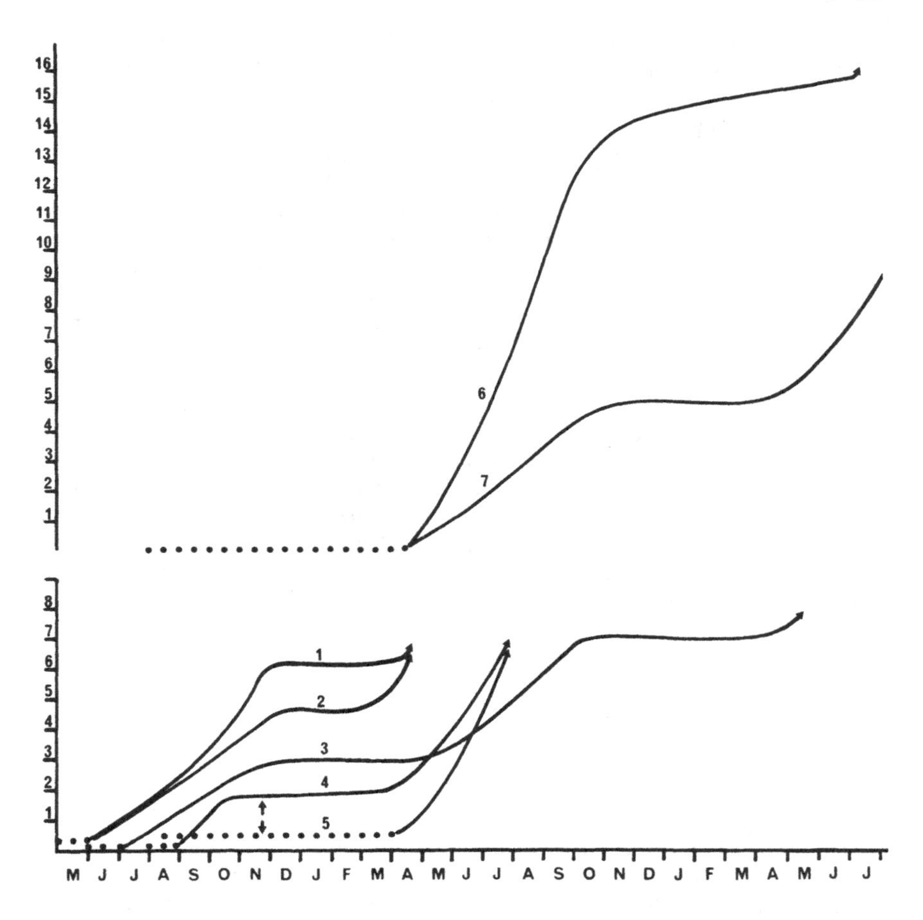

Fig. 17. Different types of life cycles in stoneflies. – 1, 2: short incubation period of egg and rapid nymphal growth; 3: short incubation period of egg and slow nymphal growth; 4: short incubation period of egg and rapid nymphal growth; 5: long incubation period of egg and rapid nymphal growth; 6: egg diapause and rapid nymphal growth; and 7: egg diapause and slow nymphal growth. The incubation period is shown by a dotted line, the nymphal growth period by a solid line.

incubation period during the winter, with hatching occurring in the spring, followed by a rapid nymphal growth period during the summer. A number of species have a two year life cycle, the egg undergoing diapause during the first winter, and the entire egg stage lasting for 8-10 months. Normally, the nymphal growth of these species mainly occurs during the first summer after hatching, the adults emerging immediately after the break-up of the ice cover the following spring. The life cycle of only a single species, *Dinocras cephalotes* regularly lasts more than two years. Ulfstrand (1969) has reported a 3 to 4-years life cycle for this species in northern Sweden. Intraspecific differences in life cycles are known, however. Brittain (1977) reported that *Nemurella pictetii* can change from a 1-year to a 2-year cycle under unfavourable environmental conditions, as often occur at high altitudes. Lillehammer (1975b, 1976) recorded a prolongation of nymphal growth from one to two years for *Nemoura cinerea,* when given an insufficient food supply. The egg stage of *Leuctra digitata* and *L. fusca* which has a short duration at high water temperatures, may last for most of the winter if eggs are deposited at low temperatures late on in the autumn (Lillehammer, 1985b). Both *Capnia atra* (Brittain *et al.,* 1984, 1986) and *Leuctra hippopus* (Lillehammer, 1976, 1987a) seem capable of giving rise to local populations with different types of life cycle.

The emergence time of some species differs widely according to the latitude (Brinck, 1949), or in different parts of a single stream (Lillehammer, 1975a). Several authors such as Dodds & Hirshaw (1925), Nebeker (1971) and Ikomonow (1973), have shown that emergence occurs later at higher altitudes. They considered that water temperature is the most important factor. Elvang & Madsen (1973), however, speculated whether the photoperiod was also a factor regulating the emergence time of *Taeniopteryx nebulosa.* Lillehammer (1975a) studied the emergence of six species of stoneflies in several streams within an area subject to the same photoperiod conditions, but where the annual temperature regimes differed widely. The emergence times of *Leuctra hippopus* and of some other species showed a delay of 3-4 months in the streams with the lowest temperatures, compared to the other streams.

Zoogeography

The plecopteran fauna of Fennoscandia and Denmark together comprises 42 species: 25 of these species have been recorded from Denmark, 35 each from Finland and Norway, and 36 from Sweden (Table 1). Most of the species are quite common, at least within their main area of distribution. A few species, however, are quite rare, viz., *Rhabdiopteryx acuminata* (two localities recorded from Finland), *Amphinemura palmeni* (recorded from only a few localities in northern Fennoscandia), *Capnia nigra* (recorded from only a few streams in south-eastern Sweden), and *Protonemura hrabei* (found in a few localities in Jutland).

Dispersal routes

The present distribution of the stonefly fauna of Fennoscandia and Denmark is the combined result of the history of postglacial dispersal, the influence of distributional barriers, and the impact of environmental factors, and the success or otherwise of the individual species in overcoming these problems in the past and at present.

The fauna includes species of southern, eastern and north-eastern origin. The species that dispersed from the south may have taken a westerly route over Denmark and southern Sweden, e.g. *Capnia bifrons* (Fig. 18), or an easterly route, e.g. *Protonemura intricata* (Fig. 19). Different species have achieved different degrees of colonization

Fig. 18. Range of *Capnia bifrons* (Newm.).

success. *Perlodes microcephala* has only managed to reach Denmark (Fig. 20), while *P. dispar* also occurs in southern Sweden and in Norway and Finland. Three other species, *Capnia bifrons, Dinocras cephalotes* (Fig. 22), *Brachyptera risi* (Fig. 21), have only managed to colonize the western part of Fennoscandia and are absent from most of Finland. The species of north-eastern origin have also shown various degree of success in colonizing Fennoscandia. A species such as *Diura nanseni* (Fig. 23) is distributed over the greater part of the area, while *Nemoura sahlbergi* (Fig. 24) has only a restricted northern distribution. The immigration route of *Leuctra digitata* (Fig. 25) is difficult to trace. This species may have reached Fennoscandia from the north-east or from the south, or even from both directions. The same may also hold true for several

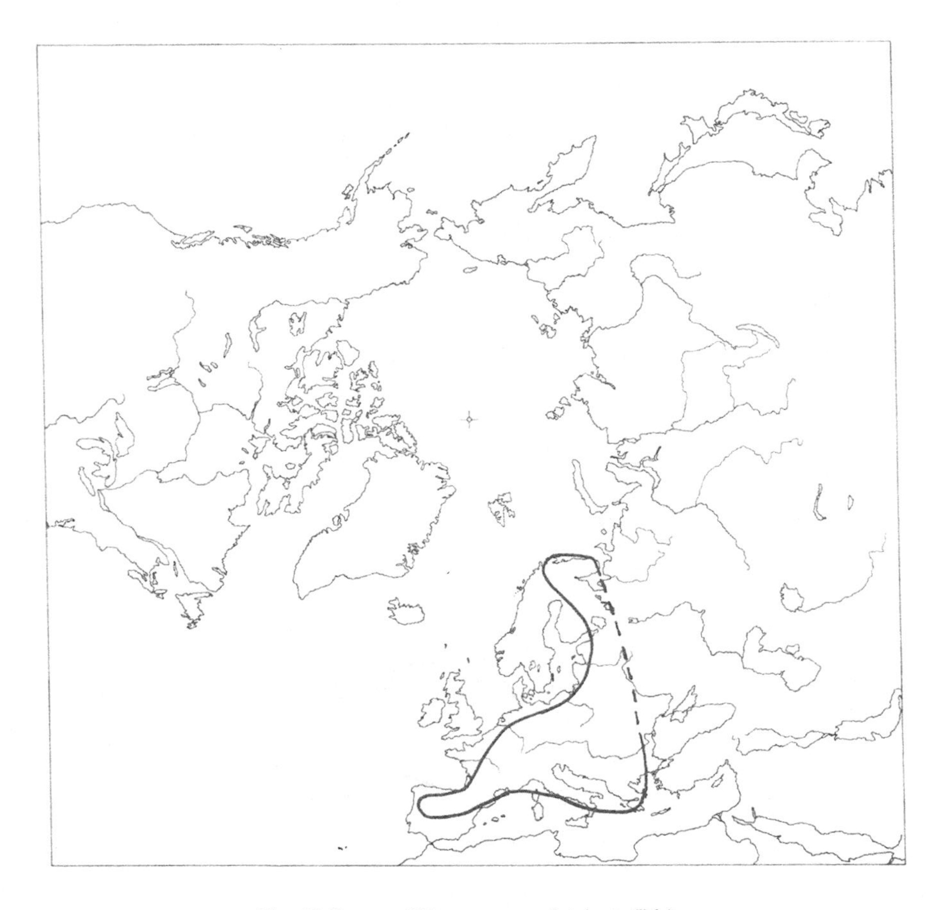

Fig. 19. Range of *Protonemura intricata* (Ris).

other species, such as *Diura bicaudata* (Fig. 26), *Leuctra hippopus* or *Amphinemura standfussi. Isoptena serricornis* (Fig. 27) must have entered Fennoscandia from the north-east (Finland and Sweden) and Denmark from the south-east.

The western-arctic faunal elements, discussed by Lindroth (1957) and Rognes (1986), and which also have counterparts in the flora (Berg, 1963; Petterson, 1983; Nordal 1985a, b) pose no problems for the plecopteran fauna of Fennoscandia. The only species with a clear eastern distribution in the Nearctic, *Diura nanseni* (Fig. 23), has a continuous distribution in the Palearctic from Norway to Kamtchatka. There is no need to postulate the former existence of a landbridge or a refugium in order to explain its present distribution.

Fig. 20. Range of *Perlodes microcephala* (Pict.).

Types of distribution

Species such as *Arcynopteryx compacta* (Fig. 28), which has an alpine-montane distribution in Europe, continuous in the north but with only relict populations in the south, *Amphinemura borealis* (Fig. 29) with a boreo-alpine distribution, and *Capnia vidua* (Fig. 30), seem to be solely represented by relict populations. In Fennoscandia *C. vidua* only occurs north of the Polar circle (Lillehammer, 1986c). The nearest other populations are found in Iceland and Scotland.

The general trend in the distribution of the stoneflies found in Fennoscandia and Denmark results from the strong influence of the species of north-eastern origin and,

Fig. 21. Range of *Brachyptera risi* (Mort.).

as a result of this, the highest number of species are found in the areas north of the Polar circle.

The strongest faunal relationships are seen between Sweden and Norway and between Finland and Norway, both of which have 84% of the species in common. The weakest relationship is seen between Denmark and Finland, with only 51% of the species in common. This is nearly the same percentage relationship as that seen between Norway and Great Britain. The wide difference seen between Finland and Denmark is an expression of the greater representation of species of north-eastern origin in Finland and of the fact that the fauna of Denmark includes a greater number of species of southern origin than the rest of Fennoscandia.

Fig. 22. Range of *Dinocras cephalotes* (Curt.).

A number of stonefly species have a circumpolar distribution, viz., *Arcynopteryx compacta* (Fig. 28), *Diura bicaudata* (Fig. 26), *Nemoura arctica* (Fig. 31) and probably also *N. sahlbergi* (Fig. 24).

Species such as *Capnia pygmaea* (Fig. 32) and *Diura nanseni* (Fig. 23) reach their southernmost distributional limits in Fennoscandia, while in *Capnia atra* isolated populations are found further south (Fig. 33). In contrast, species such as *Dinocras cephalotes* (Fig. 22) and *Brachyptera risi* (Fig. 21) have their northernmost distributional limits in the same area. Areas with a continental climate have been found to hold a larger number of species than areas with an oceanic climate (Lillehammer, 1974b, 1985a). The largest number of families and genera are also represented in the continen-

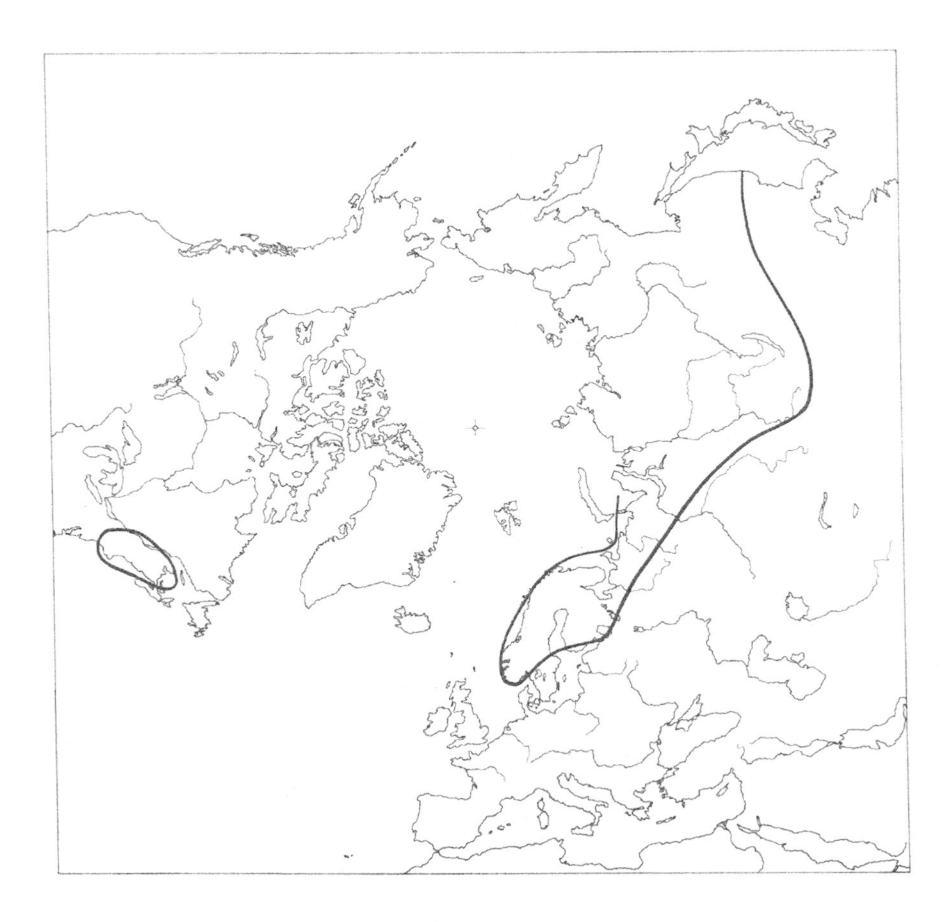

Fig. 23. Range of *Diura nanseni* (Kempny).

tal areas, the lowest in coastal areas.

The vertical distribution of the stonefly fauna has been studied in Norway and Sweden. Wide differences exist in the altitudinal distributions of the different species (Lillehammer, 1984). The number of families, genera and species represented at high altitudes is lower than at low altitudes (Lillehammer, 1985a). In the sub-alpine vegetation belt a large number of closely related species occurs, while in the mid-alpine zone only a few, less closely related, species are present. Herbivorous species predominate in the sub-alpine zone and below, while predatory species are predominant in the mid-alpine zone. Those species occurring at high altitudes generally show special adaptations to the environment, either during the egg stage or the nymphal stages (Lilleham-

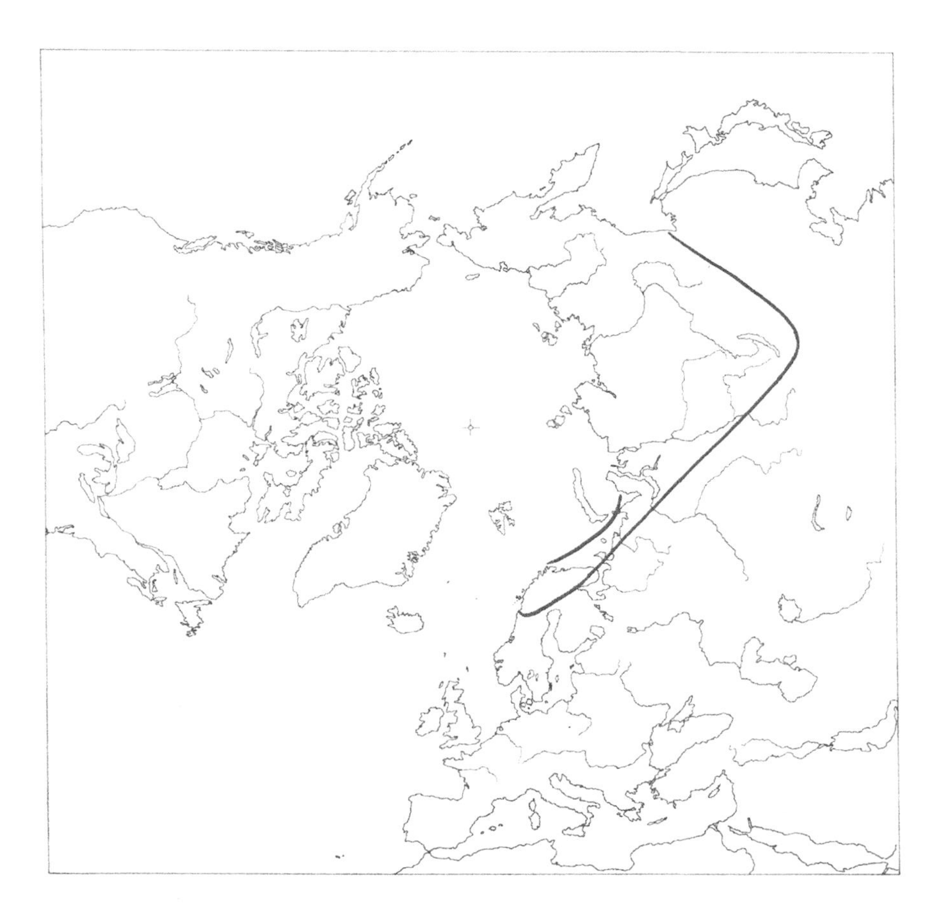

Fig. 24. Range of *Nemoura sahlbergi* Mort.

mer, 1985c; Lillehammer *et al.,* 1988).

The plecopteran fauna of a number of Fennoscandian islands has been studied by Brinck (1949) and Kaiser (1972). Relatively few species were recorded and predatory species were absent from all the islands except Funen in Denmark. This may be connected with both climatic and ecological factors, such as mentioned by Pianka (1981).

Fig. 25. Range of *Leuctra digitata* Kempny.

Phylogeny and classification

The phylogeny of the Plecoptera has been discussed by several authors. Illies (1962) has discussed the evolution of the plecopteran central nerveous system in relation to phylogeny. More recently, the evolution and the phylogenetic system of the Plecoptera has been discussed by Zwick (1973, 1974).

The stoneflies that occur in the northern hemisphere belong to the suborder Arctoperlaria, which consists of two main groups. Firstly, the Systellognatha, sub-divided into the families Pteronarcidae and Peltoperlidae and the superfamily Subulipalpia,

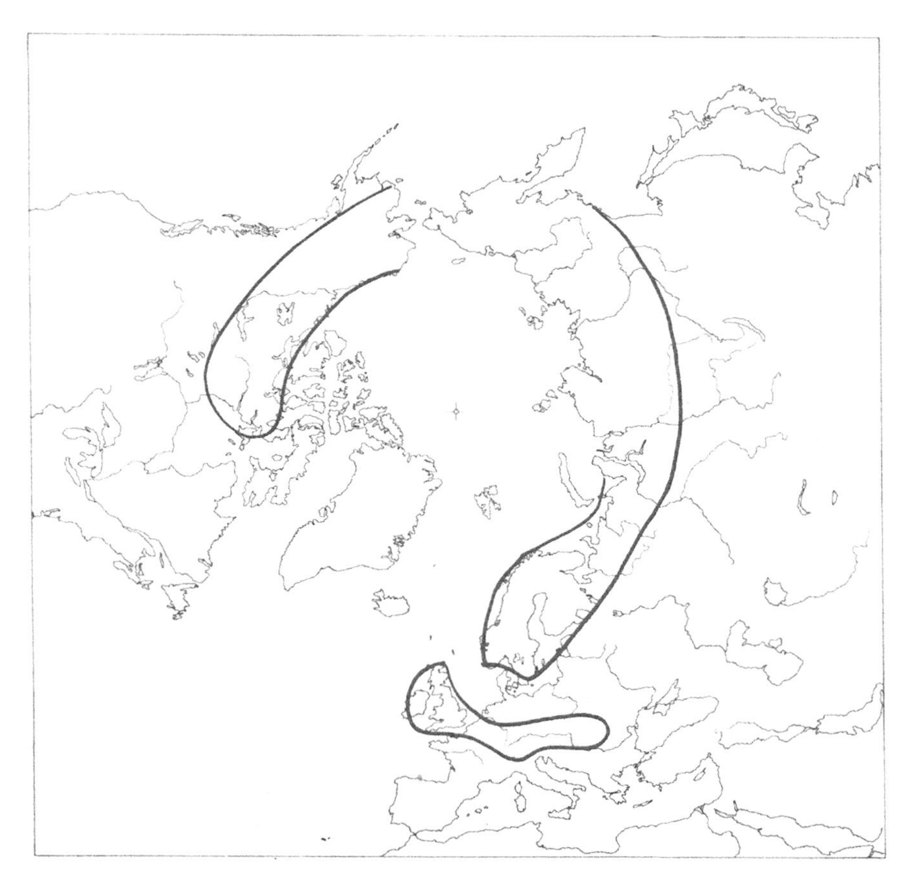

Fig. 26. Range of *Diura bicaudata* (L.).

and secondly the Euholognatha, sub-divided into the superfamilies Scopuroidea and Nemouroidea (Zwick, 1973).

The Fennoscandian stonefly fauna comprises members of the superfamily Subulipalpia, with the families Perlodidae, Perlidae and Chloroperlidae, and the superfamily Nemouroidea, with the families Taeniopterygidae, Nemouridae, Capniidae and Leuctridae (see Table 1).

The Subulipalpia have a number of synapomorphic characters: the predatory larvae possess long mandibles, the lacinia has strong teeth, the glossa is reduced, the eggs have micropyles, the gut has vesicles and the first tarsal segment is small. In Subulipalpia species the egg opening, for hatching, is situated opposite to the anchor plate,

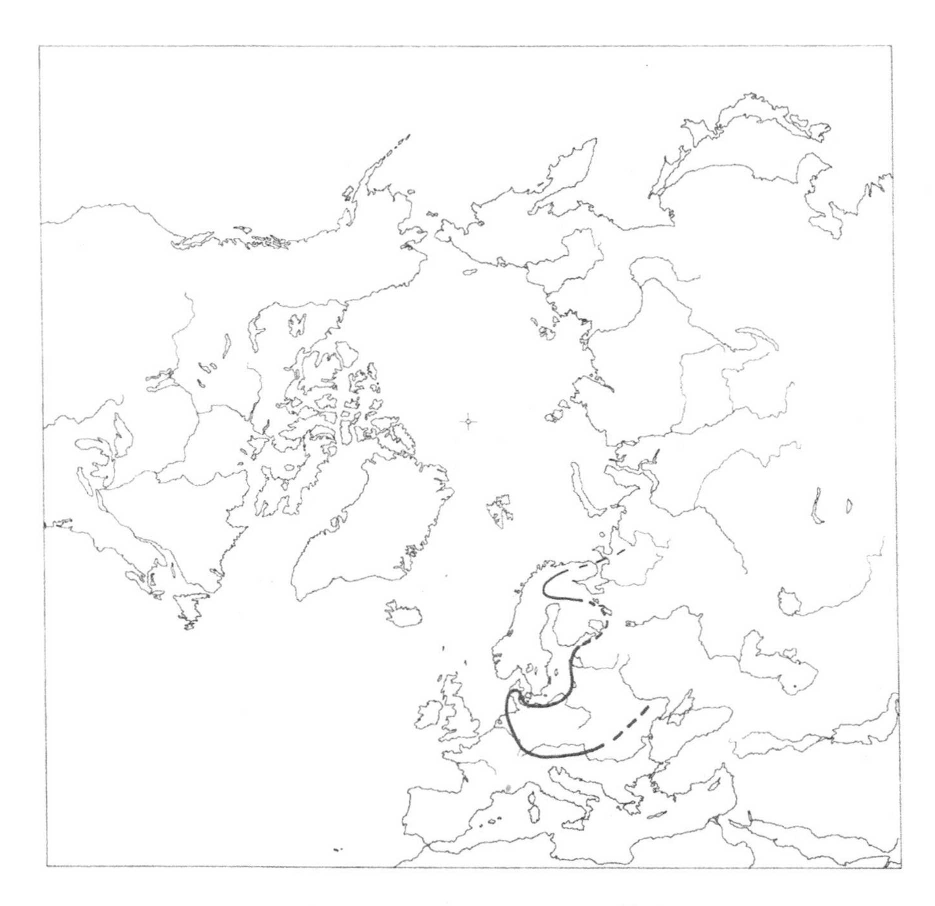

Fig. 27. Range of *Isoptera serricornis* (Pict.).

while in the Pteronarcidae and Peltoperlidae the opening lies closer to the anchor plate.

The Nemouroidea share two synapomorphic characters: the form of the mesosternum, the spina being joined to the furca, and the inner structures of the male epiproct, this being divided into an outer and an inner lobe. The phylogenetic system of the Nemouroidea has been worked out by Illies (1965).

Nelson (1984) has made numerical cladistic analyses of the phylogenetic relationships of the Plecoptera, and Bronsky (1982) has discussed the evolution of the wing structure. Stark & Szczytko (1982) have discussed the phylogeny of the Pteronarcidae on the basis of the egg morphology studied by SEM.

Fig. 28. Range of *Arcynopteryx compacta* (McL.).

Variability and evolution

Several authors, e.g. Kuhtreiber (1934), Hynes (1941), Brinck (1949), have mentioned the wide variation found in many taxonomic characters of stonefly species. Lillehammer (1974a) systematized the degree of variability and came to the conclusion that several of the characters hitherto used by taxonomists are invalied. One such character is the Nemouridae X-cross in the wing venation, previously used for separating the Nemouridae from the Capniidae plus the Leuctridae. Lillehammer (1974a) found that this character was absent in up to 50% of the members of some populations of *Am-*

Fig. 29. Range of *Amphinemura borealis* (Mort.).

phinemura and *Nemoura* species, and present to nearly the same extent in some *Capnia* species. Some of the results of this variability study are shown in Figs 135-137. A wide variation in the wing venation pattern is also seen in the Perlodidae. It is thus of little use for distinguishing the genus *Arcynopteryx* from that of *Perlodes* by the number of cross-veins present in the anterior cubital area (Fig. 55). A wide degree of variation in the characters of the genitalia of one or both sexes is also found in some species, such as *Leuctra fusca, L. hippopus, Capnia atra, C. vidua, C. bifrons, Nemoura arctica, Diura bicaudata* and *D. nanseni* (Lillehammer, 1974a).

The morphological variation of *Leuctra hippopus* in particular has been studied by Lillehammer (1976, 1986a). This species has the ability to give rise to local populations

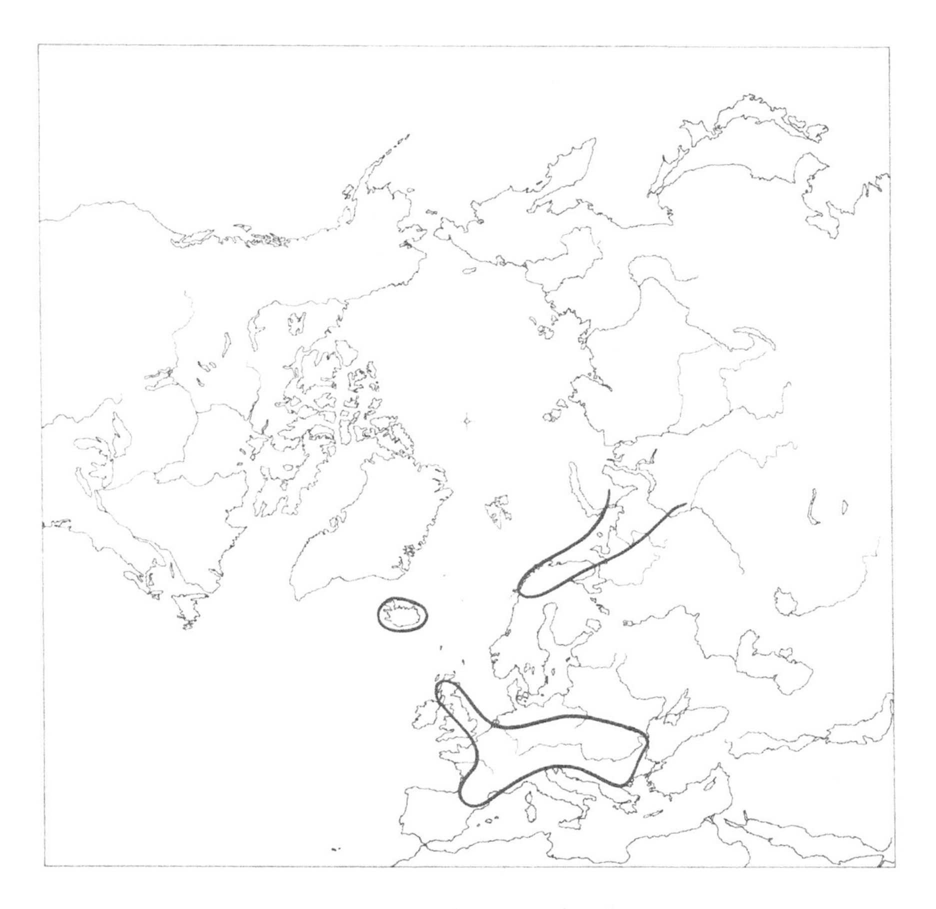

Fig. 30. Range of *Capnia vidua* Klap.

that differ from one another in several characters, such as the genitalia including the microstructures, body size, femur length, and size of the first instar nymph. Recent studies (Lillehammer, 1987a) have shown that they also differ in growth rate and type of life cycle. One such isolated population, occurring in a short stream connecting two lakes, emerges, copulates, and deposits its eggs much earlier than any other populations living in the surrounding areas (Lillehammer, 1976). Such variation is an integral part of evolution. Tauber & Tauber (1981) state that all widely distributed insect species encounter a great diversity in climatic conditions and that this is reflected by the wide variety of the seasonal life cycle found among populations living in different geographical areas.

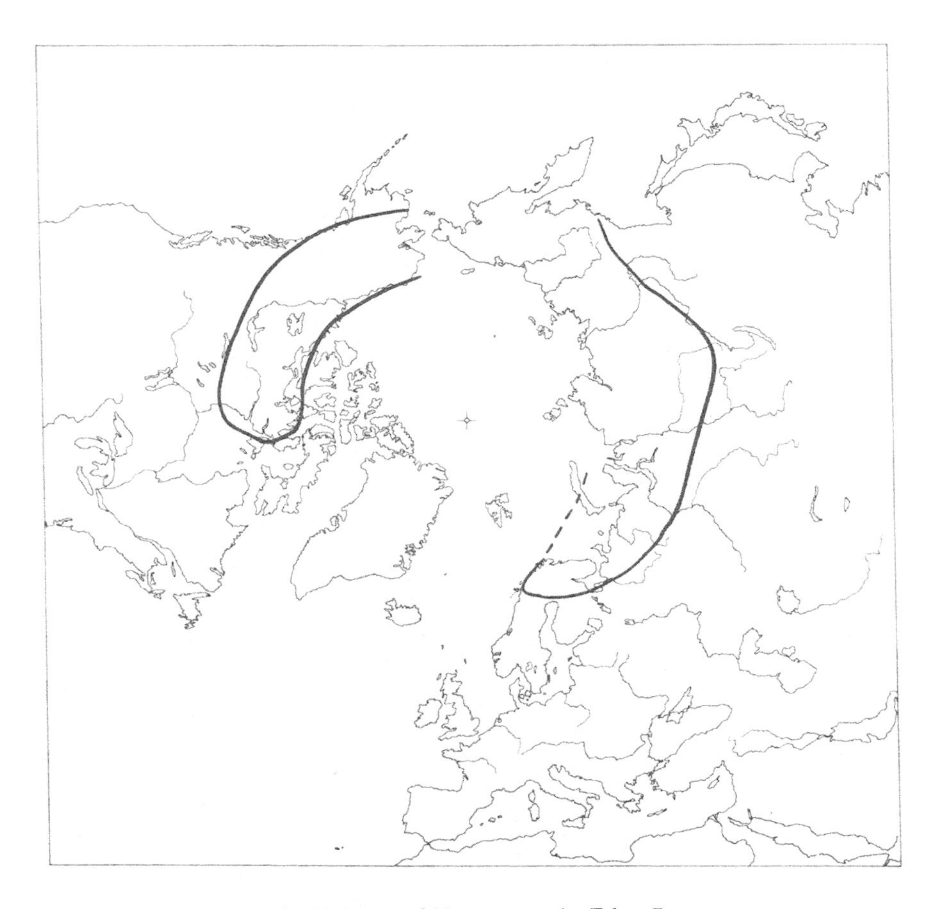

Fig. 31. Range of *Nemoura arctica* Esben-P.

In Central Europe, the result of such evolution and speciation can be seen in the separation from the *Leuctra hippopus*-group of two distinct species, *Leuctra pseudohippopus* and *L. hippopoides* (Zwick, 1973). This probably took place during or after the last glaciation. A simple ability to give rise to local population that differ from each other taxonomically and ecologically, is seen in Fennoscandia in *L. hippopus* (Lillehammer, 1974a, 1976, 1986a, 1987a) and also in *Capnia atra* (Brittain *et al.*, 1984, 1986).

Several factors in combination are at work in the process of speciation. Tauber & Tauber (1981) discuss the role of differentiation in the life cycle in diversification and speciation. The diversity, evolution and distribution of subspecies of *Capnopsis*

Fig. 32. Range of *Capnia pygmaea* (Zett.).

schilleri have been discussed by Zwick (1984). Evolution seems still to be in progress among various members of the plecopteran fauna of Fennoscandia. Evolutionary studies, based on drumming patterns and their "dialects", combined with studies of egg structure and genetical analysis using electrophoretic methods, appear to be relevant fields for further research on the stoneflies of Fennoscandia.

Sampling and preserving

Plecopteran nymphs can be sampled in streams and lakes during most of the year. During wintertime the ice cover of streams and lakes may prove difficult to penetrate. A

Fig. 33. Range of *Capnia atra* Mort.

variety of sampling methods have been used: when sampling for life cycle studies alone, or for faunistic recording, the simple kick method can be used (see Frost *et al.* 1970). For studies of abundance, a quantitative method, such as the Surber sampler, can be used. Adults can be sampled by sweeping the vegetation growing along rivers, streams and lakes, or by searching underneath stones along riversides and lake shores. Different species are present as adults from March until October. A number of emergence traps have also been used; such are described by Kuusela & Pulkkinen (1978), Brittain & Lillehammer (1978), Kuusela (1984) and Huusko & Kuusela (1985).

Both adults and numphs are best preserved in 80% alcohol. Dry-pinned specimens are of little value. Dried specimens can be softened up by warming for a few minutes in lactic acid. Species determinations are usually made under a stereomicroscope, although in some cases the body parts, mounted separately, must be examined at a higher power of magnification.

The arrangement of the bristles on the legs, pronotum or cerci, is often used in specific determinations of nymphs. Sometimes the unpigmented, thin bristles may be difficult to observe against a black or a white background. The black may absorb too much light, and the white may give too little contrast. A transparent, blue-coloured, piece of glass placed over the white background is useful in order to increase the degree of contrast.

The use of a light source will heat up the alcohol preservative and cause turbulence and movement of the object under study. To prevent this, nymphs or adults can be stabilised by first melting some paraffin wax to form a thin layer on the bottom of a small petri dish and then pin the specimen into this in the required position.

Rearing technique

The adults of most species can easily by kept under laboratory conditions. They can most successfully be collected shortly after emergence, taken back to the laboratory and kept in plastic boxes, adding twigs, leaves and water from their home stream. In general, they should be kept at 10-12°C. To hasten the process of egg maturation, the adults may be kept at temperatures of about 20°C during daytime.

Petri dishes filled with water should be provided if studies of egg incubation are to be made. When they are fully mature, the females seek water and deposit their eggs. These can then be kept at different water temperatures and the duration of the incubation period in each case can then be recorded and a regression analysis made of the results. If the incubation period lasts for a long period, additional water must be supplied at intervals.

The growth of the nymphs can also studied in the laboratory, either in running water, or in aquaria placed in refrigerators. In the latter case it is important that more water is added at intervals, or that the water is regularly changed. Most of the nymphs of herbivorous species can be reared successfully if twigs, leaves, and detritus containing micro-organisms are placed on a thin layer of sand and gravel. This must be done about a week or two before the nymphs are introduced. The Systellognatha species are carnivorous and require a continuous supply of suitable prey animals in order to grow successfully. They are therefore difficult to rear for more than the few first instar stages, during which they often feed on detritus and microorganisms.

Acknowledgements

I am grateful for help from several persons and institutions during the preparation of this book.

A very special thank is due to Mai Britt Ringdal who prepared nearly all drawings. During the entire work I have had the best cooperation with her.

I am also indebted to Karsten Sund for valuable help, to Grethe Garfjeld for typewriting the manuscript and to Philip Tallantire for correcting the language, to my colleagues Jan E. Raastad and Per Pethon for valuable discussions on systematical and zoogeographical problems, and to John E. Brittain and Svein J. Saltveit for cooperation in the ecological studies of stoneflies.

I also wish to thank the following persons for kind loan of material and for information: Lita Greve Jensen, Zoological Museum, Bergen; John Solem, D.K.N.V.S. Museum, University of Trondheim; Zoologisk avdeling, Tromsø Museum; Frank Jensen, Naturhistorisk Museum, Århus, Denmark; Kalevi Kuusela, Oulu, Finland; Pekka Hiilivirta, Zoological Museum, Helsinki, Finland; Heikki Hamalainen, University of Joensuu, Finland; Søren Langemark, Zoologisk Museum, København, Denmark; Roy Danielsson, Museum of Zoology and Entomology, Lund, Sweden; Peter Zwick, Limnologische Flusstation des Max Plank-Instituts für Limnologie, Schlitz, West Germany.

I wish to thank R. Ruprecht, Gutenberg University, Mainz for giving permission to reproduce the oscillograms published in Oikos (1972).

I am also grateful to Professor P. Brinck who in the very beginning introduced me to stoneflies, and who gave me many ideas for my later works.

Table 1. Classification of the Plecoptera of Fennoscandia and Denmark, with indication of the distribution (DK = Denmark, S = Sweden, N = Norway, SF = Finland).

Suborder Arctoperlaria
    Group Systellognatha
        Superfam. Subulipalpia
        Fam. Perlodidae
        Subfam. Perlodinae
            Genus *Arcynopteryx* Klapálek, 1904
                A. compacta (McLachlan, 1872) - S, N, SF

Genus *Diura* Billberg, 1820
D. bicaudata (Linnaeus, 1758) - S, N, SF
D. nanseni (Kempny, 1900) - S, N, SF
Genus *Isogenus* Newman, 1833
I. nubecula Newman, 1833 - DK, S, N, SF
Genus *Perlodes* Banks, 1903
P. dispar (Rambur, 1842) - DK, S, N, SF
P. microcephala (Pictet, 1833) - DK
Subfam. Isoperlinae
Genus *Isoperla* Banks, 1906
I. difformis (Klapálek, 1909) - DK, S, N, SF
I. grammatica (Poda, 1761) - DK, S, N, SF
I. obscura (Zetterstedt, 1840) - S, N, SF
Fam. Perlidae
Subfam. Perlinae
Genus *Dinocras* Klapálek, 1907
D. cephalotes (Curtis, 1827) - DK, S, N
Fam. Chloroperlidae
Subfam. Chloroperlinae
Genus *Isoptena* Enderlein, 1909
I. serricornis (Pictet, 1841) - DK, S, SF
Genus *Siphonoperla* Zwick, 1967
S. burmeisteri (Pictet, 1841) - DK, S, N, SF
Genus *Xanthoperla* Zwick, 1967
X. apicalis Newman, 1836 - S, N, SF

Group Euholognatha
Superfam. Nemouroidea
Fam. Taeniopterygidae
Subfam. Taeniopteryginae
Genus *Taeniopteryx* Pictet, 1841
T. nebulosa (Linnaeus, 1758) - DK, S, N, SF
Subfam. Brachypterinae
Genus *Brachyptera* Newport, 1849
B. braueri (Klapálek, 1900) - DK, S
B. risi (Morton, 1896) - DK, S, N, SF
Genus *Rhabdiopteryx* Klapálek, 1902
R. acuminata Klapálek, 1905 - SF
Fam. Nemouridae
Genus *Amphinemura* Ris, 1902
A. borealis (Morton, 1894) - S, N, SF
A. palmeni Koponen, 1916 - N, SF
A. standfussi (Ris, 1902) - DK, S, N, SF
A. sulcicollis (Stephens, 1836) - DK, S, N, SF

Genus *Nemoura* Latreille, 1796
N. arctica Esben-Petersen, 1910 - S, N, SF
N. avicularis Morton, 1894 - DK, S, N, SF
N. cinerea (Retzius, 1783) - DK, S, N, SF
N. dubitans Morton, 1894 - DK, S, SF
N. flexuosa Aubert, 1949 - DK, S, N, SF
N. sahlbergi Morton, 1896 - S, N, SF
N. viki Lillehammer, 1972 - N, SF
Genus *Nemurella* Kempny, 1898
N. pictetii Klapálek, 1900 - DK, S, N, SF
Genus *Protonemura* Kempny, 1898
P. hrabei Raušer, 1956 - DK
P. intricata (Ris, 1902) - N, SF
P. meyeri (Pictet, 1841) - DK, S, N, SF
Fam. Capniidae
Genus *Capnia* Pictet, 1841
C. atra Morton, 1896 - S, N, SF
C. bifrons (Newman, 1839) - DK, S, N
C. nigra (Pictet, 1833) - S
C. pygmaea (Zetterstedt, 1840) - S, N, SF
C. vidua Klapálek, 1904 - S, N, SF
Genus *Capnopsis* Morton, 1896
C. schilleri (Rostock, 1892) - S, N, SF
Fam. Leuctridae
Subfam. Leuctrinae
Genus *Leuctra* Stephens, 1836
L. digitata Kempny, 1899 - DK, S, N, SF
L. fusca (Linnaeus, 1758) - DK, S, N, SF
L. hippopus Kempny, 1899 - DK, S, N, SF
L. nigra (Olivier, 1811) - DK, S, N, SF

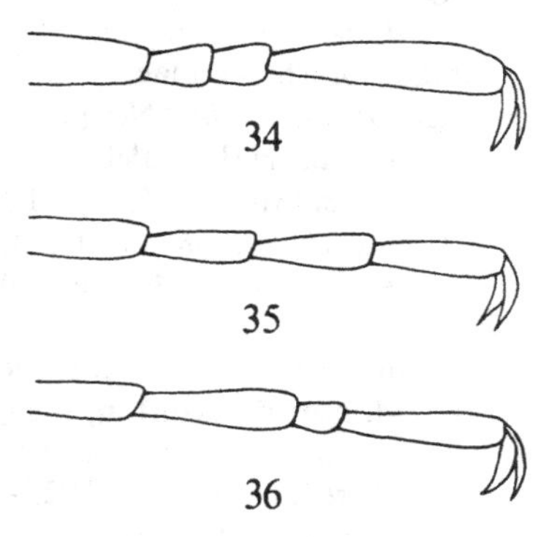

Figs 34-36. Tarsi in 34: Perlodidae, Perlidae and Chloroperlidae. 35: Taeniopterygidae; 36: Nemouridae, Leuctridae and Capniidae.

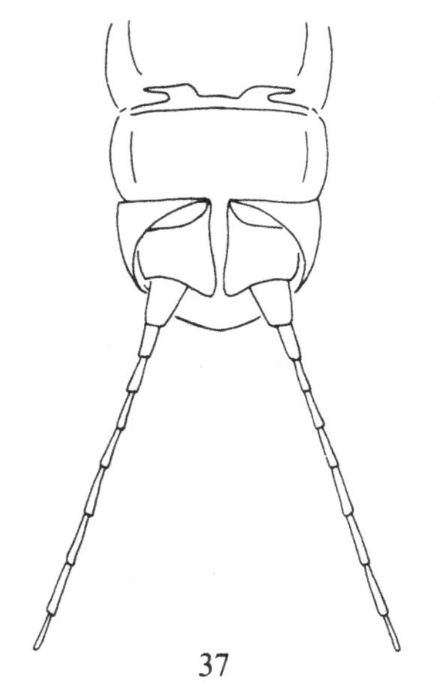

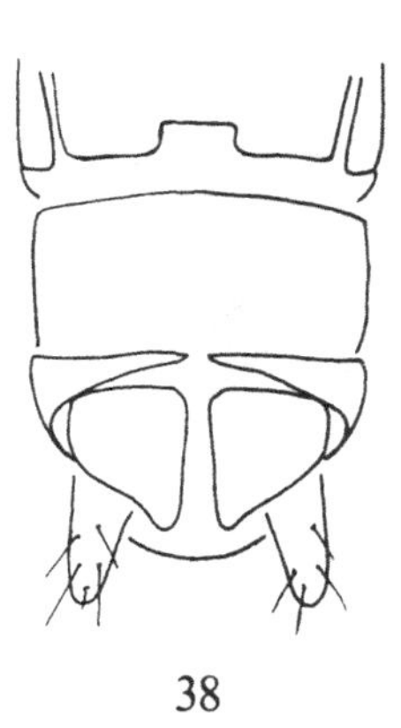

Figs 37, 38. Apex of adult abdomen showing multi-segmented cerci in 37: Capniidae, and one-segmented cerci in 38: Leuctridae and Nemouridae.

Key to families of Plecoptera

Adults

1	Third tarsal segment longer than first and second segments together (Fig. 34)	6
–	Third tarsal segment shorter than first and second segments together (Figs 35, 36)	2
2(1)	Each tarsal segment longer than preceding segment (Fig. 35)	Taeniopterygidae (p. 85)
–	Second tarsal segment much shorter than both first and third segments, which are of subequal length (Fig. 36)	3
3(2)	Full-winged specimens	4
–	Short-winged males	Capniidae (p. 126)
4(3)	Cubital area without cross-veins (Fig. 39). Cerci with several segments (Fig. 37)	Capniidae (p. 126)
–	Cubital area with several cross-veins (Fig. 40). Cerci with one segment (Fig. 38)	5
5(4)	Veins Scl and Sc2 present on both wings (Fig. 40)	Nemouridae (p. 89)
–	Only Scl present on both wings (Fig. 41)	Leuctridae (p. 140)
6(1)	Full-winged specimens	7
–	Short-winged males	9

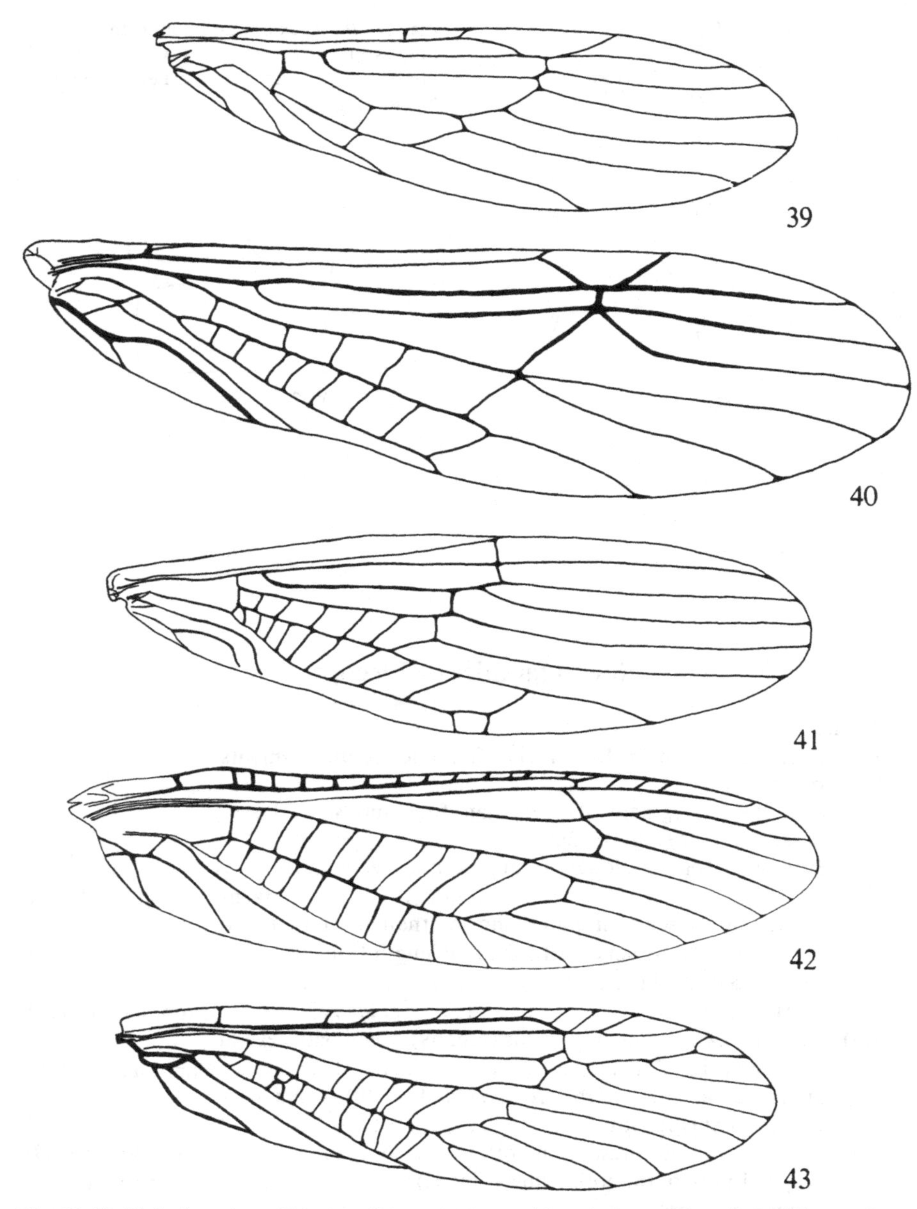

Figs 39-43. Right forewing of 39: Capniidae; 40: Nemouridae; 41: Leuctridae; 42: Perlidae; and 43: Perlodidae.

7(6) Anal area of hindwing small, all anal veins simple (Fig. 9) .... Chloroperlidae (p. 72)
- Anal area of hindwing large, some anal veins branched (Fig. 10) 8
8(7) Cross-vein r-m of forewing arises from R4+5 (Fig. 42)....... Perlidae (p. 71)
- Cross-vein r-m of forewing arises from Rs before this divides (Fig. 43) .... Perlodidae (p. 49)
9(6) Numerous cross-veins present between costa and subcosta. Wing veins irregular (Fig. 11) .... Perlidae (p. 71)
- Few cross-veins present between costa and subcosta. Wing veins often regular (Fig. 12) .... Perlodidae (p. 49)

Nymphs

1 Glossa as long as paraglossa (Fig. 44). Mandibles short and stout 2
- Glossa reduced and much smaller than paraglossa (Fig. 45). Mandibles elongate 5
2(1) Each tarsal segment as long as or longer than preceding segment (Fig. 35) .... Taeniopterygidae (p. 85)
- Second tarsal segment shorter than first segment (Fig. 36)............... 3
3(2) Body shape cylindrical and elongate. Hind leg, when stretched alongside abdomen, not reaching abdominal apex (Fig. 54).. 4
- Body shape stout. Hind leg, when stretched alongside abdomen, greatly over-reaching abdominal apex (Fig. 53) .. Nemouridae (p. 89)

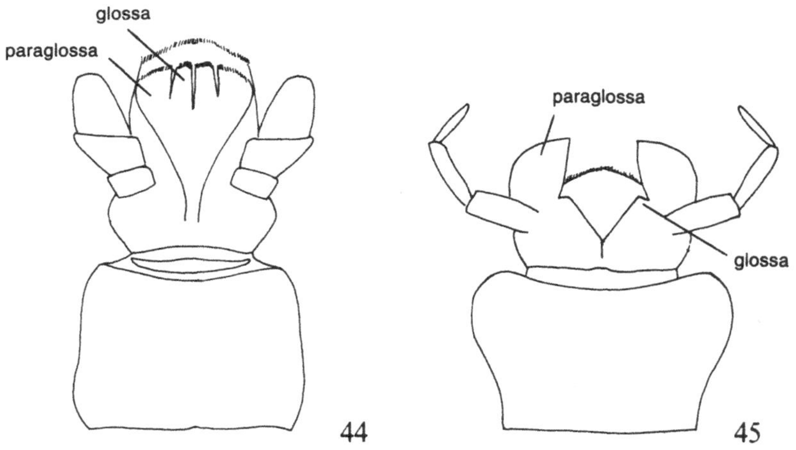

Figs 44, 45. Labium in ventral view of 44: euholognath nymph and 45: systellognath nymph.

4(3) Abdominal segments 1-4 with discrete tergites and sternites (Fig. 48). Paraproct longer than wide. Successive cercal segments rapidly increasing in length (Fig. 50) .... Leuctridae (p. 140)
– Abdominal segments 1-9 with discrete tergites and sternites (Fig. 49). Paraproct wider than long. Successive cercal segments slowly increasing in length (Fig. 51) .... Capniidae (p. 126)
5(1) Thorax with pleural gills laterally (Fig. 52) .... Perlidae (p. 71)
– Thorax without pleural gills 6

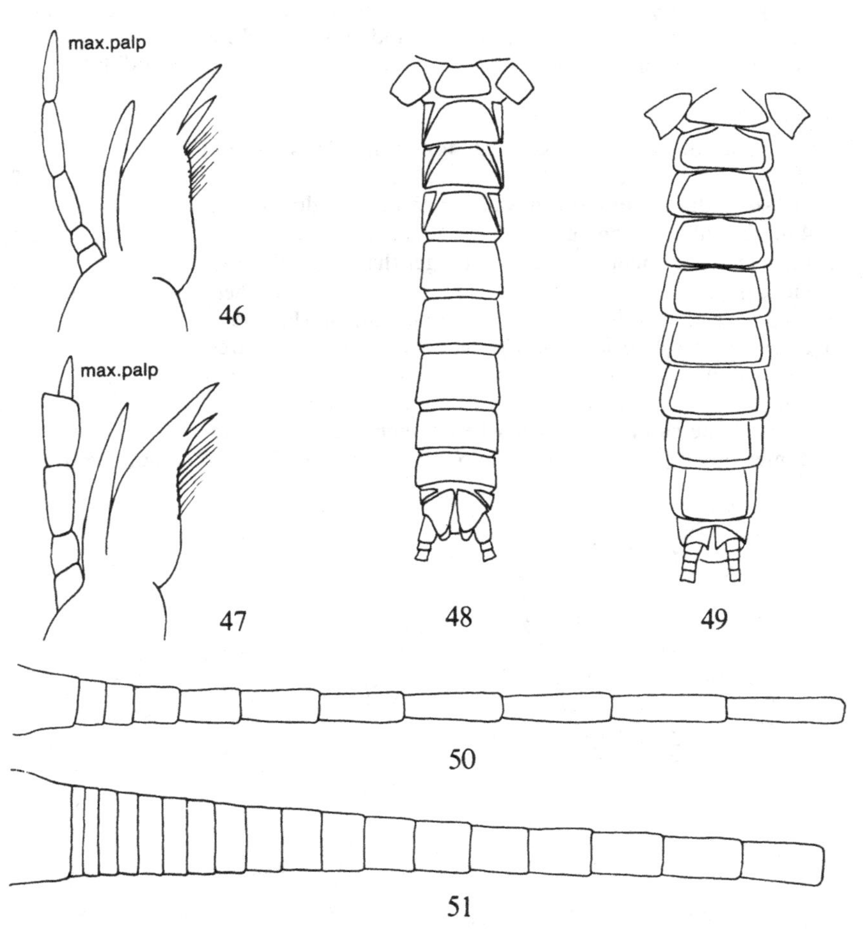

Figs 46, 47. Maxilla with maxillar palp in 46: Perlodidae and 47: Chloroperlidae.
Figs 48, 49. Abdomen in ventral view in 48: Leuctridae and 49: Capniidae.
Figs 50, 51. Cercus of nymphs in 50: Leuctridae and 51: Capniidae.

6(5) Last segment of maxillary palp reduced, only about one-quarter as wide as width of preceding segment (Fig. 47) .... Chloroperlidae (p. 72)

– Last segment of maxillary palp normal, i.e. more than one-quarter as wide as width of preceding segment (Fig. 46) .... Perlodidae (p. 49)

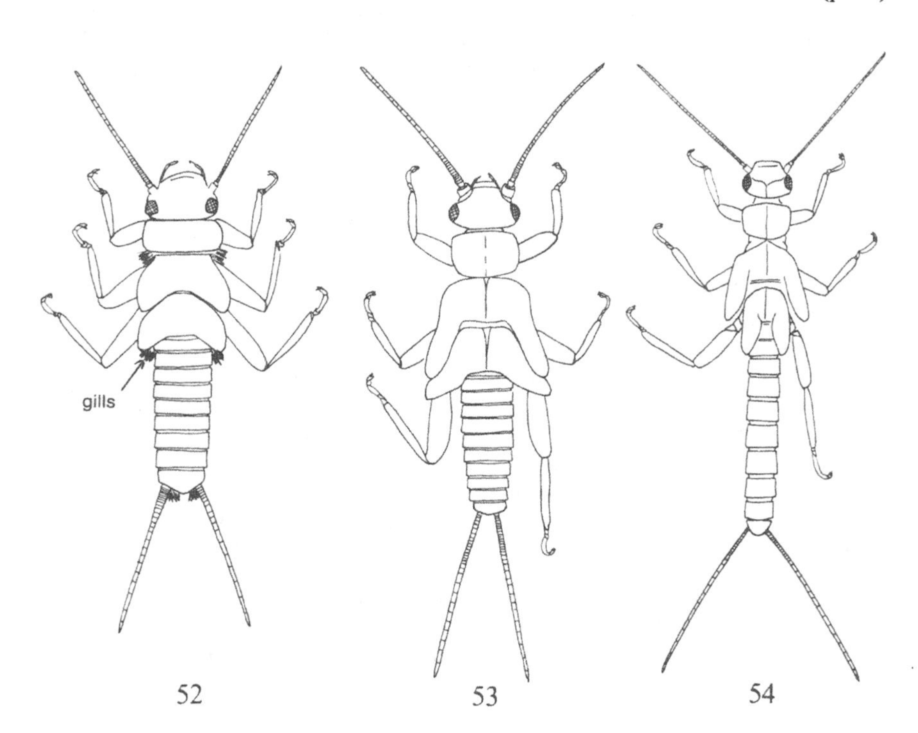

Figs 52-54. Nymphs of 52: *Dinocras cephalotes* (Curt.); 53: Nemouridae; and 54: Leuctridae or Capniidae. The arrow shows the position of the filamentous gills.

Family Perlodidae

The family shares one character with the Chloroperlidae: the cross-vein r-m arises from R before this divides (Figs 8, 42); in Perlidae r-m arises from R4+5. The anal area of the hindwing is well developed with some branched veins; in Chloroperlidae the anal area is reduced and the veins are simple (Fig. 9). Perlodidae females are usually full-winged, while the males of some species are obligatory micropterous. Most of the species are dark brown in colour, but deep green species occur. Usually the male is smaller than the female.

The family is divided into two subfamilies: Perlodinae and Isoperlinae. While the Perlodinae is a heterogeneous group, the Isoperlinae is a homogeneous group, the males of which can be separated from those of the Perlodinae by their bent, finger-shaped paraproct and possession of sclerotised teeth on the penis. Some authors, e.g. Kimmins (1950), Brinck (1952), Illies (1955), and Hynes (1977), use the shape of vein R2+3 in separating the two subfamilies. They state that R2+3 is forked in the Perlodinae and unforked in the Isoperlinae. Lillehammer (1974a) showed that intermediate forms occur, and therefore this character alone cannot be used.

Key to species of Perlodidae

Adults

1	Full-winged specimens, males or females	2
–	Short-winged males (Fig. 12)	11
2(1)	Forewing with several irregular cross-veins at apex (Fig. 55)	3
–	Forewing without irregular cross-veins at apex (Fig. 43)	6
3(2)	Males . . . . 6. *Perlodes microcephala* (Pictet)	
–	Females	4
4(3)	Subgenital plate simple (Figs 63, 64)	5
–	Subgenital plate bilobed (Fig. 62) . . . . 1. *Arcynopteryx compacta* (McLachlan)	
5(3)	Subgenital plate large, covering most of segment 9 and without incision at middle (Fig. 63) . . . . 5. *Perlodes dispar* (Rambur)	
–	Subgenital plate covering about half of segment 9; a low incision at middle (Fig. 64) . . . . 6. *Perlodes microcephala* (Pictet)	
6(2)	Vein R2+3 forked. Two or more cross-veins between C and R1	7
–	Vein R2+3 simple. One, seldom two, cross-veins between C and R1 . . .	9

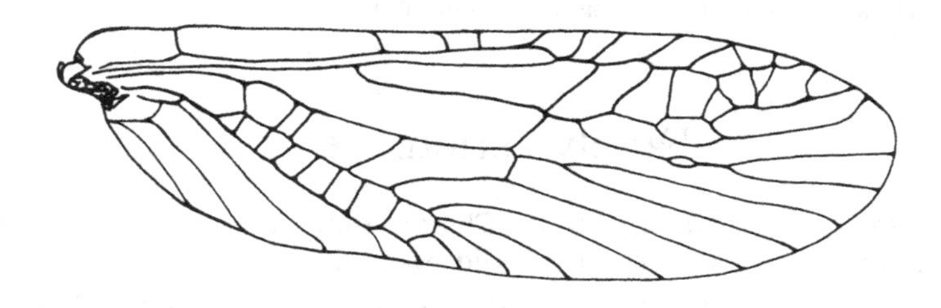

Fig. 55. Forewing of *Arcynopteryx compacta* (McL.).

7(6) Tergite 10 of male longitudinally divided (Fig. 59). Female subgenital plate large, covering most of the segment (Fig. 65) .... 4. *Isogenus nubecula* Newman

\- Tergite 10 of male undivided. Female subgenital plate covering about half of the segment (Figs 66, 67) 8

8(7) Male short-winged. Subanal lobe simple (Fig. 60). Female subgenital plate small, incurved or straight anteriorly (Fig. 66)............................ 2. *Diura bicaudata* (Linnaeus)

\- Male usually full-winged, occasionally short-winged. Subanal lobe thicker in the middle (Fig. 61). Female subgenital plate small, rounded anteriorly (Fig. 67)... 3. *Diura nanseni* (Kempny)

9(6) Female full-winged, male short-winged. Female subgenital plate large, rectangular (Fig. 71) ..... 7. *Isoperla difformis* (Klapálek)

\- Both sexes full-winged 10

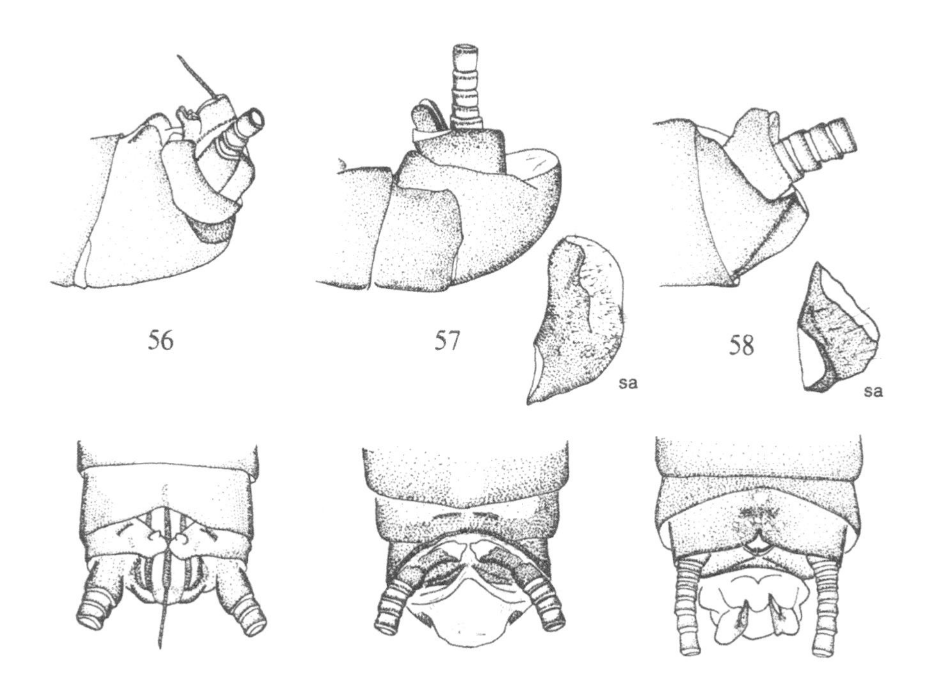

Figs 56-58. Male abdominal apex of Perlodidae, upper row in lateral view, lower row in dorsal view; sa = subanal lobe. - 56: *Arcynopteryx compacta* (McL.); 57: *Perlodes dispar* (Ramb.); 58: *P. microcephala* (Pict.).

10(9) Penial armature (pa) visible from below (Fig. 69). Female full-winged, subgenital plate rounded posteriorly (Fig. 72) .... 8. *Isoperla grammatica* (Poda)
- Penial armature (pa) invisible (Fig. 70). Female full-winged, subgenital plate pointed posteriorly (Fig. 73) .... 9. *Isoperla obscura* (Zetterstedt)

11(1) Sternum 8 with a ventral lobe. (Fig. 68, vl) 7. *Isoperla difformis* (Klapálek)
- Sternum 8 without a ventral lobe 12

12(11) Tergite 10 undivided (Fig. 57) 13
- Tergite 10 divided (Fig. 56) ..... 1. *Arcynopteryx compacta* (McLachlan)

13(12) Forewings at apex with several cross-veins (Fig. 55) .... 5. *Perlodes dispar* (Rambur)
- Forewings without irregular cross-veins 14

14(13) Epiproct simple (Fig. 60) .... 2. *Diura bicaudata* (Linnaeus)
- Epiproct thickened in the middle (Fig. 61) ... 3. *Diura nanseni* (Kempny)

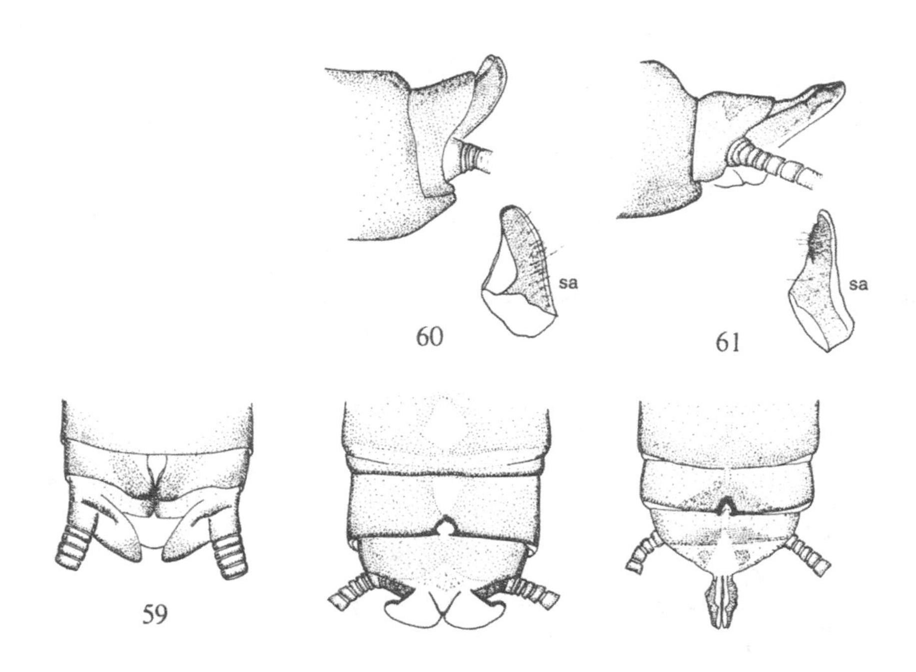

Figs 59-61. Male abdominal apex of Perlodidae, upper row in lateral view, lower row in dorsal view; sa = subanal lobe. - 59: *Isogenus nubecula* Newm., 60: *Diura bicaudata* (L.); 61: *D. nanseni* (Kempny).

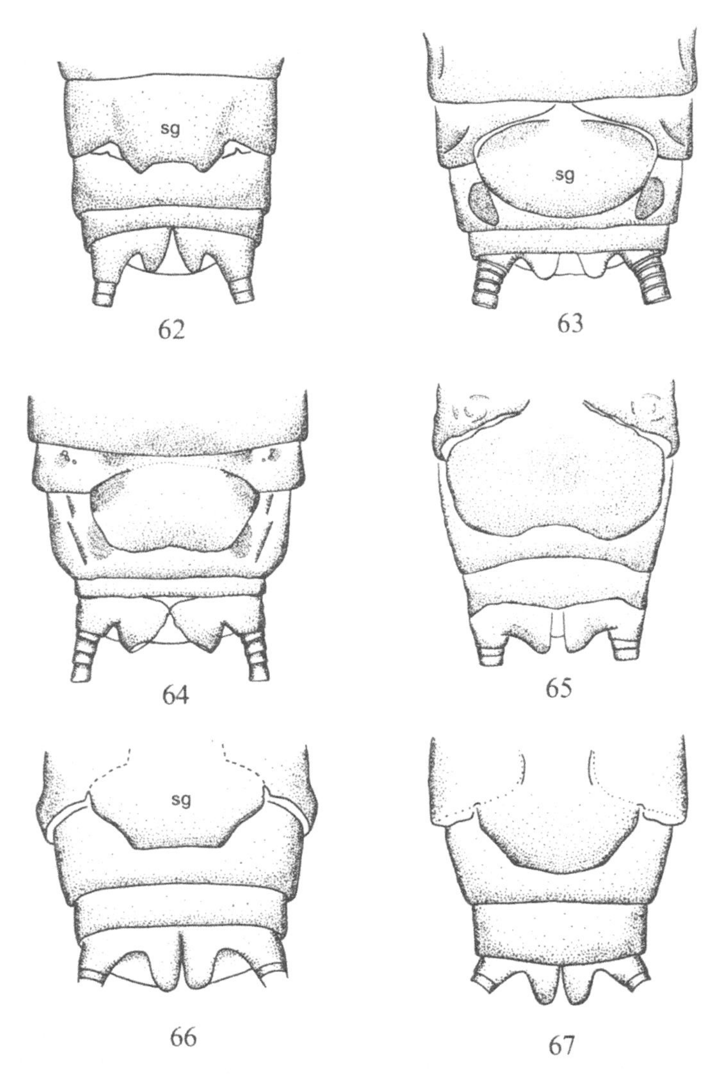

Figs 62-67. Female abdominal apex of Perlodidae, ventral view; sg = subgenital plate. - 62: *Arcynopteryx compacta* (McL.); 63: *Perlodes dispar* (Ramb.); 64: *P. microcephala* (Pict.); 65: *Isogenus nubecula* Newm.; 66: *Diura bicaudata* (L.); 67: *D. nanseni* (Kempny).

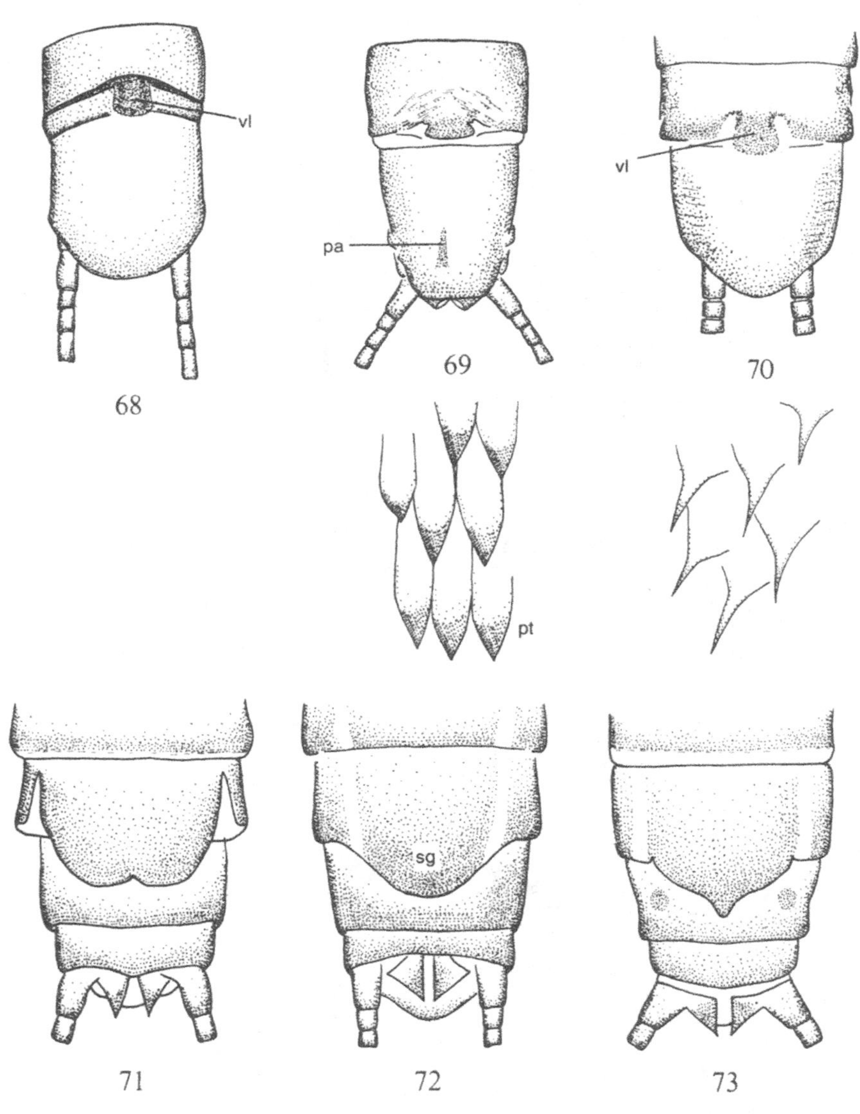

Figs 68-70. Male abdominal apex of *Isoperla* spp., ventral view; vl = ventral lobe, pa = penial armature, pt = penial teeth. - 68: *Isoperla difformis* (Klap.); 69: *I. grammatica* (Poda); 70: *I. obscura* (Zett.).

Figs 71-73. Female abdominal apex of *Isoperla* spp., ventral view; sg = subgenital plate. - 71: *Isoperla difformis* (Klap.); 72: *I. grammatica* (Poda); 73: *I. obscura* (Zett.).

Nymphs

1 Abdominal segments 1-4 with discrete tergite and sternite (Fig. 76). Lacinia evenly narrowing towards apex (Figs 80, 81) 2
- Abdominal segments 1-3 with discrete tergite and sternite (Fig. 74). Subapical tooth on inner margin of lacinia incerted at an angle (Fig. 77) ..... 1. *Arcynopteryx compacta* (McLachlan)
- Abdominal segments 1 and 2 with discrete tergite and sternite (Fig. 75) 3

2(1) Body dark brown. Head without distinct lighter parts in front of M-line and between ocelli (Fig. 86) . 5. *Perlodes dispar* (Rambur)

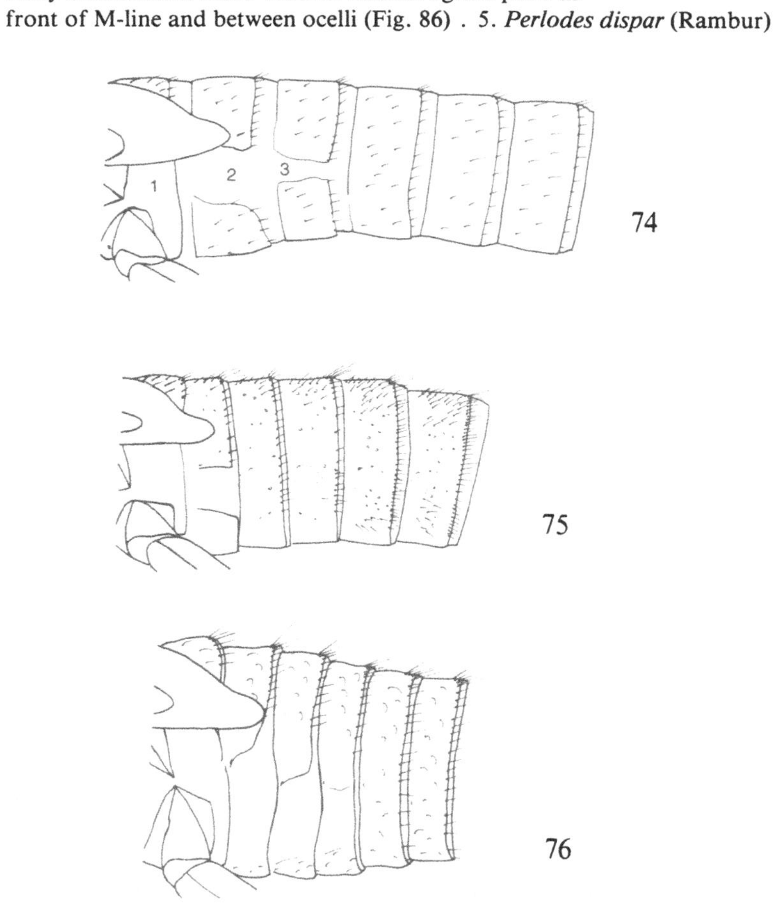

Figs 74-76. First abdominal segments of nymphs of Perlodidae; lateral view. - 74: *Arcynopteryx compacta* (McL.); 75: *Diura nanseni* (Kempny); and 76: *Perlodes dispar* (Ramb.).

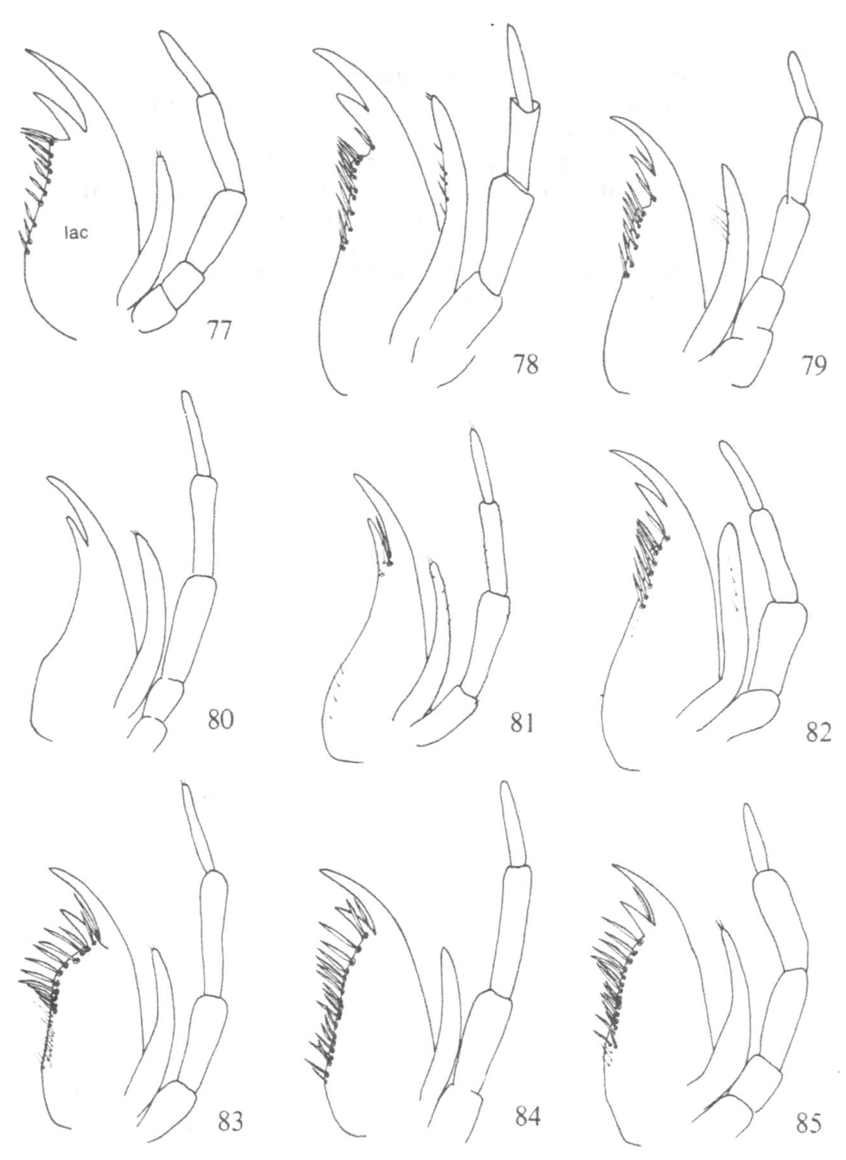

Figs 77-85. Maxilla of nymphs of Perlodidae; lac = lacinia. - 77: *Arcynopteryx compacta* (McL.); 78: *Diura bicaudata* (L.); 79: *D. nanseni* (Kempny); 80: *Perlodes dispar* (Ramb.); 81: *P. microcephala* (Pict.); 82: *Isogenus nubecula* Newm.; 83: *Isoperla difformis* (Klap.); 84: *I. grammatica* (Poda); and 85: *I. obscura* (Zett.).

- Body light brown with darker patches. Head with distinct lighter parts in front of M-line and between ocelli (Fig. 87) .... 6. *Perlodes microcephala* (Pictet)

3(1) Lacinia emarginate just below subapical tooth; with wide gap to fringe of bristles on inner margin (Figs 78, 79) 4

- Inner margin of lacinia only slightly notched below subapical tooth, with only a narrow gap to fringe of bristles on inner margin (Figs 82-85) 5

4(3) Pronotum with dense fringe composed of short stout bristles only (Fig. 88).................. 2. *Diura bicaudata* (Linnaeus)

- Pronotum with fringe composed of both long and short bristles (Fig. 89) .... 3. *Diura nanseni* (Kempny)

5(3) Body with scattered clothing hairs (Fig. 90). Lacinia as in Fig. 82 .... 4. *Isogenus nubecula* Newman

- Body with dense black clothing hairs (Fig. 91) 6

6(5) Cercal segments 7-10 with both long and short bristles posteriorly, and with long intermediate bristles scattered over the entire segment (Fig. 97). Head light pigmented anteriorly (Fig. 92) .... 7. *Isoperla difformis* (Klapálek)

- Cercal segments 7-10 with only short bristles posteriorly, with a fringe of intermediate bristles laterally on segments 15-17 (Figs 98, 99) 7

7(6) Cercal segments 15-17 with dense fringe of intermediate bristles laterally (Fig. 99). Head with light patches around eyes (Fig. 93) .... 9. *Isoperla obscura* (Zetterstedt)

- Cercal segments 15-17 with open fringe of bristles (Fig. 98). Head without light patches around eyes (Fig. 94).................. .... 8. *Isoperla grammatica* (Poda)

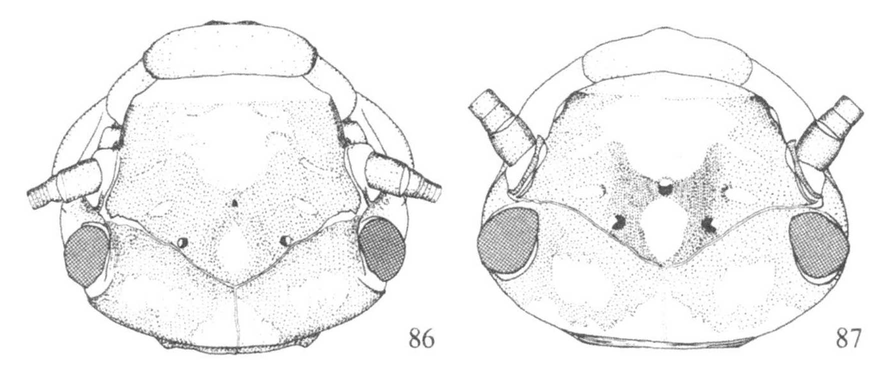

Figs 86, 87. Head of nymphs of 86: *Perlodes dispar* (Ramb.) and 87: *P. microcephala* (Pict.).

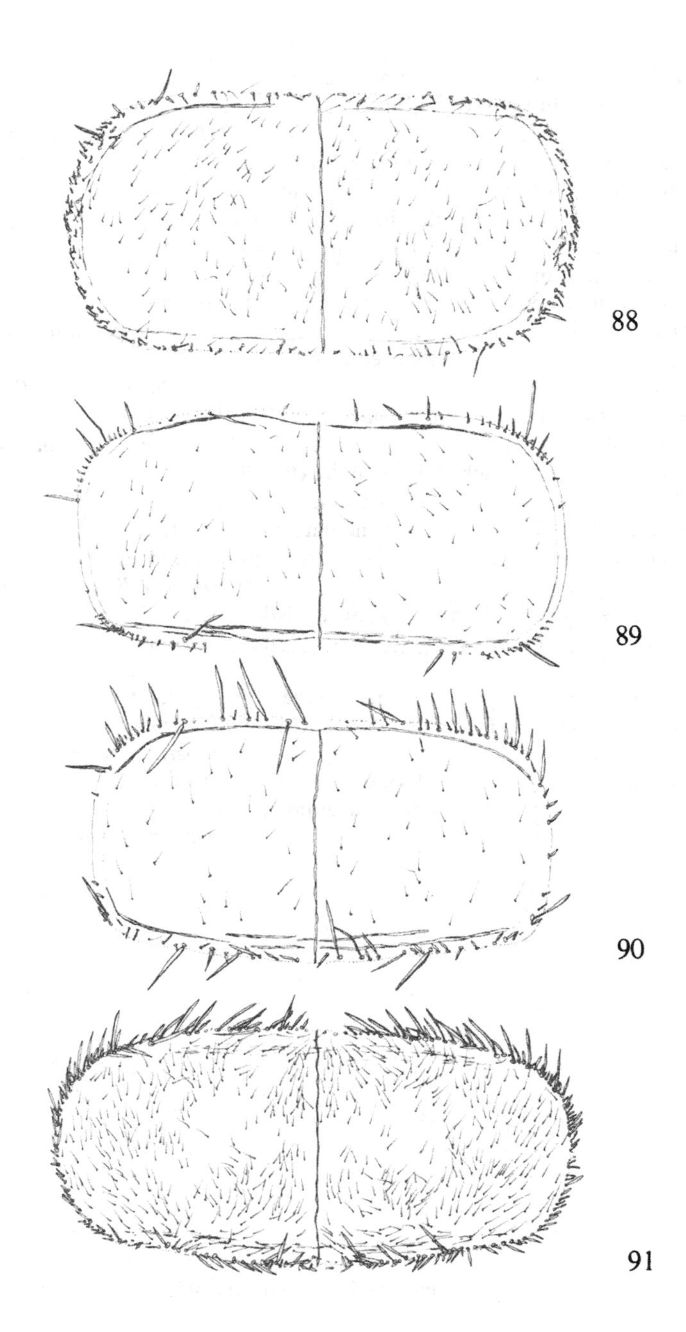
88
89
90
91

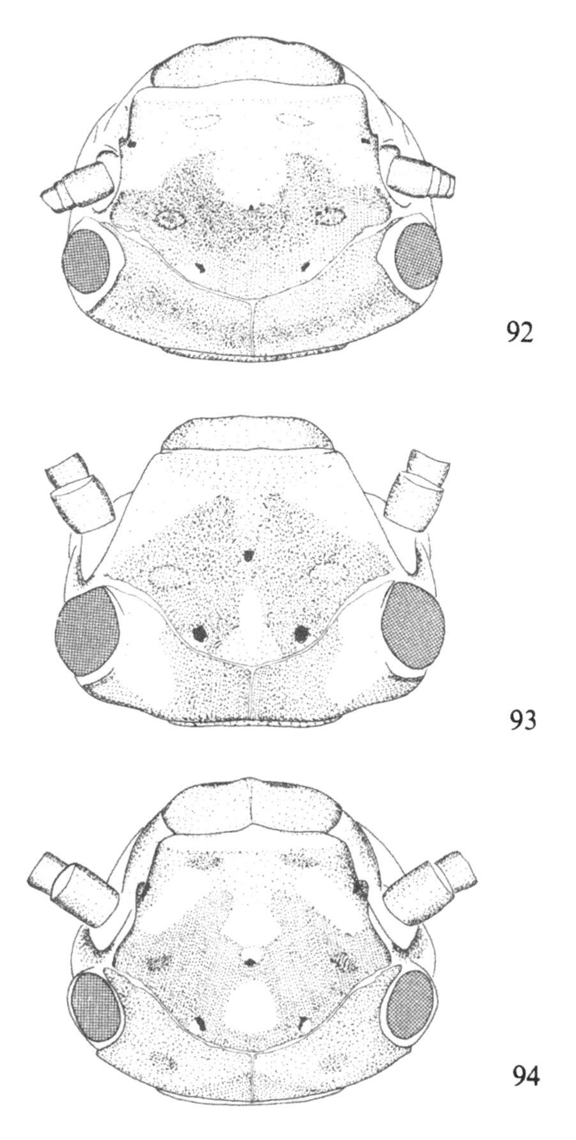

Figs 92-94. Head of nymphs of 92: *Isoperla difformis* (Klap.); 93: *I. obscura* (Zett.); and 94: *I. grammatica* (Poda).

Figs 88-91. Bristle arrangement on nymphal pronotum in Perlodidae. - 88: *Diura bicaudata* (L.); 89: *D. nanseni* (Kempny); 90: *Isogenus nubecula* Newm.; 91: *Isoperla obscura* (Zett.).

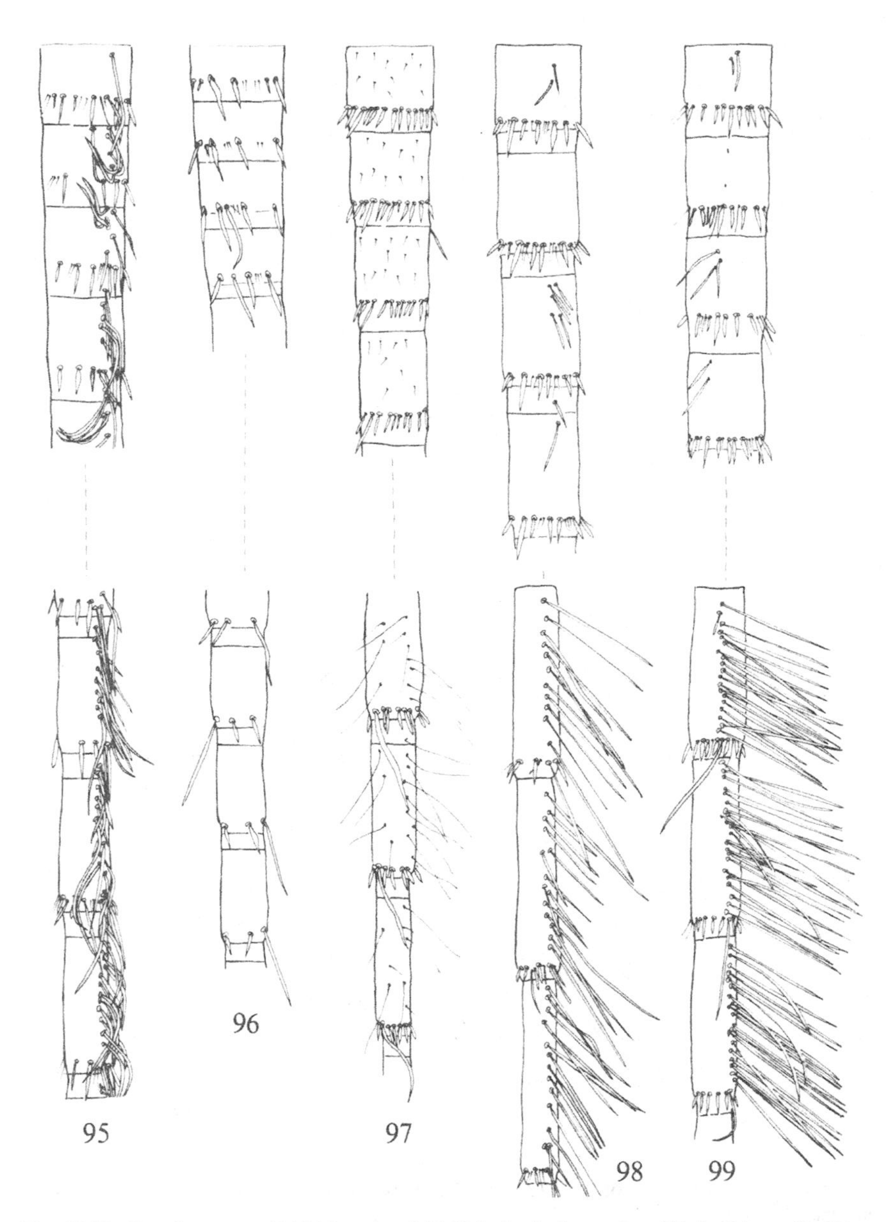

Figs 95-99. Cercal segments 7-10 (above) and 15-17 (below) of nymphs of Perlodidae. - 95: *Diura bicaudata* (L.); 96: *Isogenus nubecula* Newm.; 97: *Isoperla difformis* (Klap.); 98: *I. grammatica* (Poda); and 99: *I. obscura* (Zett.).

SUBFAMILY PERLODINAE

Arcynopteryx and *Perlodes* are separated from *Diura* and *Isogenus* by the irregularity of the venation at the apex of the forewings (Figs 43, 45). *Arcynopteryx* males differ from those of *Perlodes* by having tergite 10 divided into two lobes. The females differ in the form of the subgenital plate. The same differences distinguish *Isogenus* from *Diura*. All the Perlodinae species are large stoneflies and the females have functional wings. The males of some species are micropterous and they do not possess any flying ability. The Perlodinae nymphs have only light coloured bristles on the body, while the bristles of the Isoperlinae are dark. All the Perlodinae species are predators with strong and well developed mouthparts.

Four genera with six species occur in Fennoscandia and Denmark.

Genus *Arcynopteryx* Klapálek, 1904

Arcynopteryx Klapálek, 1904b, Bull. int. Acad. Sci. Bohème, 9: 7.
Type species: *Dictyopteryx compacta* McLachlan, 1872 (des. Klapálek, 1912).

Large, dark-pigmented stoneflies with light brown wings that look old and worn. Wing venation varies a great deal, most often being irregular and with several cross-veins at apex. The full-winged female is much larger than the micropterous male. The cerci are often yellow. Males have characteristic genital appendages, the 10th abdominal tergite being divided into two lobes (Fig. 56). They also possess a cowl, a deep pouch in tergite 10. The sclerotised parts of the cowl is often used for identification of species. Nymphs are large, dark green-grey in colour. They can easily be distinguished on the characteristic form of the lacinia (Fig. 77). Abdominal segments 1-3 with discrete tergite and sternite (Fig. 74).

Only one species occurs in Europe.

1. *Arcynopteryx compacta* (McLachlan, 1872)
Figs 10, 28, 55, 56, 62, 74, 77.

Dictyopteryx compacta McLachlan, 1872, Annls Soc. ent. Belg., 15: 53.
Dictyopteryx dichroa McLachlan, 1872, Annls Soc. ent. Belg., 15: 52.
Arcynopteryx dovrensis Morton, 1901, Entomologist's mon. Mag., 37: 146.
Perlodes slossonae Banks, 1914, Proc. Acad. nat. Sci. Philad., 66: 608.
Arcynopteryx lineata Smith, 1917, Trans. Am. ent. Soc., 43: 476.
Arcynopteryx Ringdahli Bengtsson, 1933, Lunds Univ. Årsskr. (2), 29: 18.
Arcynopteryx brachifer Bengtsson, 1933, Lunds Univ. Årsskr. (2), 29: 21.

Male. Body length 10.5-16.5 mm, wing length 2.7-5.2 mm (micropterous). Head and thorax dark brown, wings light brown, wing veins darker and variable in pattern. Head broader than pronotum. Antennae long and stout, yellow at the base. Pronotum has a yellow-brown mid-band. Tergite 9 with a median incision, tergite 10 divided into two lobes which point upwards (Fig. 56). The male abdominal apex is dominated by a cowl, a deep pouch in tergite 10. A characteristic stylet is fastened to the bottom of the cowl. Cerci long and stout, the basal segments often yellow.

Female. Much larger than the male, body length 12.0-21.0 mm, wing length 6.2-18.5 mm, usually full-winged. Body colour same as in male. Subgenital plate, which covers the genital opening, has two lateral lobes (Fig. 62).

Nymphs. First instar nymph 1.2-1.7 mm. Fully grown male nymphs 14-17 mm, female nymphs 18-26 mm. Yellow-brown to light grey-brown or brown in colour. A characteristic M-line on the head. Inner margin of lacinia distinctly angled at the subapical tooth (Fig. 77). Pronotal margin with a sparse fringe of short bristles. Abdominal segments 1-3 with discrete tergite and sternite (Fig. 74). Fully grown male nymphs can be identified on the stylus, now visible on segment 10.

Variability. Some variation occurs in the shape of the male genital appendages, especially in the shape of the sclerotised parts of the cowl. There is also a large variation in body length of both sexes. In females this variation is largest in populations from the central mountains of southern Norway. A wide variation is found in the shape of the female subgenital plate (Lillehammer, 1974a). Also the wing length can vary considerably. The males are always micropterous. The females of some populations may have long wings while those of others populations have short wings.

Distribution. *A. compacta* has a circumpolar distribution, and occur in northern Finland, Norway and Sweden. In southern Norway as well as in continental Europe it only occurs at high altitudes. This species is absent from Denmark, as also from Great Britain, but occurs in the alpine regions of southern Germany and the Pyrenees (Fig. 28).

Biology. The nymph is usually a predator on other freshwater insect larvae and nymphs, but at certain times of the year it is omnivorous, consuming large quantities of plant fragments as well. *A. compacta* has a two-year life cycle: the first winter is spent in an egg diapause, hatching taking place the following spring (Lillehammer, 1985a). Growth is very rapid during the summer and the nymph is nearly full-grown by the start of the second winter. In the south, *A. compacta* mainly occurs in streams and rivers within and above the subalpine vegetation zone. In northern Fennoscandia the species also occurs in lakes. Flight period from June to August, depending on altitude and latitude.

Genus *Diura* Billberg, 1820

Diura Billberg, 1820, Enum. Ins. Mus. Billberg: 96.

Type species: *Phryganea bicaudata* Linnaeus, 1758 (mon.).

Large and dark-pigmented stoneflies, wings light brown with darker veins. Females larger than males. Head usually as wide as pronotum. Antenna long and slender. Males with characteristic subanal lobes (Figs 60, 61). Females have a subgenital plate that varies much in shape. Wing venation usually without irregular cross-veins at apex. However, the venation shows a strong variation. Males of *Diura* have tergite 10 undivided as males of *Perlodes,* but *Diura* can be separated from *Perlodes* by the regularity of its wing venation.

Two species occur in Europe. One species has always micropterous males, the other occasionally. Nymphs grey-green to dark grey. Head with a light dorsal M-line, lacinia with an incision below the subapical tooth (Figs 78, 79). Pronotum with light bristles. Abdominal segments 1 and 2 have discrete tergite and sternite (Fig. 75). The wing veins af both species are highly variable, and cross-veins between R1, R2+3, R4+5 may occur. There is also a strong tendency to forking in the median area and reduction in the cubital area.

2. ***Diura bicaudata*** (Linnaeus, 1758)
Figs 12, 26, 60, 66, 78, 88, 95.

Pryganea bicaudata Linnaeus, 1758, Syst. Nat., 10 Ed., 1: 908.
Perla postica Walker, 1852, Cat. Neur. Ins. Brit. Mus., 1: 144.
Dictyopteryx norvegica Kempny, 1852, Verh. zool.-bot. Ges. Wien, 50: 86.
Dictyopteryx septentrionis Klapálek, 1904c, Rozpr. české Acad. Cís. Fr. Jos., II, 13: 4.
Dictyopterygella parva Koponen, 1915, Meddn Soc. Fauna Flora fenn., 41: 25.

Male. Body length 9.2-17.0 mm. Wing micropterous, 2.7-4.3 mm in length, and light brown with darker veins. Head as wide as pronotum, dark brown with yellow patches. Body dark brown with short grey-brown bristles. Pronotum with a yellow mid-band. Abdomen well sclerotised and dark brown. Terga 9 and 10 undivided, subanal lobes cylindrical, outer side sclerotised, with short bristles scattered over the surface (Fig. 60).

Female. Body length 11.0-21.0 mm. Wings usually of full length, light brown with darker veins, wing length 7.8-17.4 mm. Head and pronotum dark brown and shiny, pronotum with a yellow mid-band. Abdominal segments lighter brown. On sternite 8 a subgenital plate of varying form (Fig. 66).

Nymphs. First instar nymph 1.1-1.5 mm. Fully grown male nymphs 9.5-17.0 mm, female nymphs 17-24 mm. Body yellow-brown to grey-brown with yellow markings dorsally. Head with a distinct M-line. Pronotum with a dense fringe of short stout bristles (Fig. 88). Lacinia (Fig. 78) with an obvious gap between the subapical tooth and the inner fringe of bristle. Cercal segments with a lateral fringe of long bristles (Fig. 95).

Variability. The species is polymorphic and shows variation in nearly all the taxonomically useful morphological characters. The variation in the shape of the female subgenital plate is particularly wide (Lillehammer, 1974a). A wide variation has also been recorded in the wing length and wing venation, especially in the females. Also the variation in body length is wide in both sexes. Ruprecht (1972) has recorded dialects in the drumming signals of specimens from widely separated populations (Fig. 16).

Distribution. *D. bicaudata* has a circumpolar distribution. It occurs in Finland,

Sweden and Norway, but not in Denmark. It is also recorded from Great Britain and Germany and may have reached Fennoscandia from both the north-east and the south-west (Fig. 26).

Biology. The nymph is predatory and feeds on benthic invertebrates, included crustaceans. In northern Fennoscandia it occurs in streams, rivers, and lakes, but in the southern part of the distributional area mainly in lakes. The species has a two-year life cycle. It has an egg diapause that lasts 8-10 months during the first winter. Hatching occurs in the following spring. The major part of the nymphal growth takes place during the summertime, but may also continue during the winter (Lillehammer, 1978a). Flight period from June to August, depending upon altitude and latitude.

3. ***Diura nanseni*** (Kempny, 1900)
Figs 1, 23, 61, 67, 75, 79, 89.

Isogenus nanseni Kempny, 1900, Verh. zool.-bot. Ges. Wien, 50: 90.
Dictyopterygella subfissa Bengtsson, 1931, K. svenska VetenskAkad. Skr. Naturskydds., 18: 58.

Male. Body length 10.0-15.0 mm, wing length 5.8-12.0 mm. The males are usually long-winged, but micropterous specimens do occur in some localities (Lillehammer, 1974a; Saltveit & Brittain, 1986). Head marked with a yellow-brown M. Body dark brown, pronotum with a yellow midband, wings light brown. Terga 9 and 10 undivided. Subanal lobes swollen in the middle and with long bristles at apex (Fig. 61).

Female. Resembles *D. bicaudata* but is more slender. Body length 11.5-18.5 mm, wing length 10.9-17.7 mm. Mainly long-winged. Head dark brown with yellow patches. Body dark brown, pronotum with a yellow mid-band. The form of the subgenital plate is usually different from that of *D. bicaudata* (Fig. 66), but intermediate forms occur (Lillehammer, 1974a).

Nymphs. First instar nymph 0.8-1.0 mm. Body length of full-grown male nymphs 11-16 mm, of female nymphs 12-18 mm. Head with yellow M-line, pronotum with relatively few bristles; these of different length (Fig. 89). Lacinia (Fig. 79) has a distinct gap between the subapical tooth and the inner fringe of bristles. Cercal segments with a lateral fringe of long bristles (as in Fig. 95).

Variability. Some variation occurs in the shape of the male genital appendages and a wide variation is found in the male wing length: from micropterous to full-winged (Raušer, 1971; Lillehammer, 1974a; Saltveit & Brittain, 1986). An extraordinary wide variation is found in the form of the female subgenital plate (Brinck, 1949; Benedetto, 1973a; Lillehammer, 1974a). Some variation also occurs in the female wing length, some specimens being brachypterous. The pattern of the veins may also vary in both sexes. Also the characters of the nymphs vary, especially the form of lacinia varies during the course of nymphal development (Saltveit, 1978).

Distribution. *D. nanseni* has a holarctic distribution. In Norway it occurs at its southernmost limit, and is the most common predatory stonefly species. The species does not occur in southern Sweden or in Denmark. The species is supposed to be a north-eastern immigrant to Fennoscandia (Fig. 23).

Biology. The nymph is predatory and feeds to a large degree on chironomid larvae. The species may at times be omnivorous. In northern Fennoscandia this species occurs in small streams as well as in large rivers and lakes. In southern Norway, *D. nanseni* is mainly recorded in streams and rivers, sometimes also in small streams that holds little water during the winter. The species has a two-year life cycle. It has an egg diapause that lasts for 8-10 months during the first winter. Hatching occurs in spring or early summer and the nymph grows rapidly during the summer and autumn. Emergence occurs in May-July the following year, and the flight period may last until August.

Genus *Isogenus* Newman, 1833

Isogenus Newman, 1833, Ent. Mag., 1: 415.
Type species: *Isogenus nubecula* Newman, 1833 (mon.).

Large dark-pigmented species. Head as wide as pronotum, antenna long and slender. Pronotum with a yellow mid-band. Wings light brown with darker veins. Venation at apex of wing regular, and wings of normal length in both sexes. Tergite 10 divided in the male (Fig. 59). Female subgenital plate markedly larger than in our other species of Perlodinae (Fig. 65). The genus contains only one species. Nymphs dark green, abdominal segments 1 and 2 with discrete tergite and sternite (as in Fig. 75). Body with light coloured bristles. Lacinia without incision proximal to the subapical tooth (Fig. 82).

4. ***Isogenus nubecula*** Newman, 1833
Figs 59, 65, 82, 90, 96.

Isogenus nubecula Newman, 1833, Ent. Mag., 1: 415.
Perla parisina Rambur, 1842, Hist. nat. Ins., Névr.: 450.
Perla proxima Rambur, 1842, Hist. nat. Ins., Névr.: 451.
Isogenus pudens Bengtsson, 1933, Lunds Univ. Årsskr. (2), 29: 16.

Male. Body length 14.0-17.0 mm, wing length 14.6-15.8 mm. Body dark brown, pronotum with a yellow mid-band. Tergite 10 divided and subanal plate stout and much broader than in *Arcynopteryx compacta* (cf. Figs 56 and 59). Below this structure are a supra-anal lobe and the retractile sclerotised stylus.

Female. Body length 15.0-17.5 mm, wing length 15.2-16.1 mm. Head and thorax brown, pronotum with a yellow mid-line, head with yellow patches, and abdominal segments lighter. The subgenital plate large, covering most of sternum 9 (Fig. 65).

Nymphs. Body length of full-grown male nymphs 14-16 mm, of female nymphs 16-21 mm. Brown to dark brown or yellow-brown in colour. Inner margin of lacinia increases evenly in width below the subapical tooth (Fig. 82). Whole body including pronotum (Fig. 90) sparsely covered by light-coloured bristles. The margin of pronotum with an open row of long bristles. Cercal segments without lateral fringe of long bristles (Fig. 96).

Variability. Little is known about the variability of this species.

Distribution. *I. nubecula* is found in Finland, Sweden, Norway and Denmark, although it does not occur in western Norway and has only once been recorded from Denmark (in WJ). The species occurs in Germany and in Central Europe, and is also recorded from Great Britain. The species is supposed to be an eastern imigrant to Fennoscandia.

Biology. The nymph is predatory, but little is so far known about the biology of this species, which occurs in streams and rivers. The species is supposed to have a one-year life cycle, adult emergence occurring in May-June. In River Glomma in Norway, one of the largest in Fennoscandia, *I. nubecula* occurs in the lower parts, while *Diura nanseni* occurs in the same type of habitat in the upper parts. The species may therefore have a biology that resembles that of *D. nanseni.*

Genus *Perlodes* Banks, 1903

Perlodes Banks, 1903, Ent. News, 14: 241.

Type species: *Perla microcephala* Pictet, 1833 (des. Banks, 1903).

Large species. Body dark brown to black, with yellow markings on the head and a yellow mid-band on pronotum. Wings light brown with darker veins which are irregularly arranged at apex. The genitalia is similar to those of *Diura* species. Nymphs grey-brown. Lacinia simple, without bristles on the inner side (Figs 80, 81). Abdominal segments 1-4 with discrete tergite and sternite (Fig. 76).

There are three species in this European genus; two of them occur in Fennoscandia and Denmark. The taxonomical differences between the two species occurring in Fennoscandia are slight, and the variations are not studied. A correct identification may therefore often be difficult.

5. *Perlodes dispar* (Rambur, 1842)

Figs 57, 63, 76, 80, 86.

Perlodes dispar Rambur, 1842, Hist. nat. Ins., Névr: 451.
Perlodes Zetterstedti Bengtsson, 1933, Lunds Univ. Årsskr. (2), 29: 9.

Male. Body length 11.0-17.0 mm, wing length 4.0-4.7 mm (micropterous). Body dark brown, head with three yellow marks posteriorly. Pronotum with a yellow mid-band. Wings light brown with darker veins. Male genitalia with simple appendages, subanal lobe rounded apically (Fig. 57).

Female. Body length 15.0-20.0 mm, wing length 14.6-19.1 mm. Colour as in male. Subgenital plate large and rounded, covering most of sternum 9, which also possesses two distinct dark areas laterally (Fig. 64).

Nymphs. Body length of full-grown male nymphs 14-18 mm, of female nymphs 18-26 mm. Body grey-brown with yellow markings. Head without a distinct M-line (Fig. 86). The head markings in this species are less distinct than in *P. microcephala* nymphs. Pronotum with a fringe of short bristles. Lacinia gradually pointed (Fig. 80). Cercal segments with a lateral fringe of long bristles (as in Fig. 95).

Variability. The range of variation has not been recorded. The stability of the characters used for identification may therefore be questioned.

Distribution. *Perlodes dispar* occurs in Denmark, southern Sweden and in the southernmost part of south-eastern Norway. The species has also been recorded from south-eastern Norway. The species has also been recorded from south-eastern Finland. In Denmark the species has only been recorded from the river Gudenå in EJ. The species occurs in N. Germany but has not been recorded from Great Britain.

Biology. The nymph is predatory and occurs mainly in large rivers, but can be found also in stony streams. In Norway it is only recorded in shallow streams with a high summer temperature. The species is supposed to have a one-year life cycle (Zwick, 1980). Adults emerge in May-June.

6. ***Perlodes microcephala*** (Pictet, 1833)
Figs 20, 57, 64, 81, 87.

Perla microcephala Pictet, 1833, Annls Sci. nat., 28: 59.
Perla rectangula Pictet, 1841, Perlides: 159.
Perla hispanica Rambur, 1842, Hist. nat. Ins., Névr.: 452.

Male. Body length 12-18 mm, wing length 7.5-11 mm. Body dark brown, wings light brown, with darker veins which are irregularly arranged and have cross-veins at apex. Head with lighter markings between the ocelli. Pronotum with a yellow mid-band. Genital appendages (Fig. 58), subanal lobe pointed apically.

Female. Body length 17-25 mm, wing length 14-18.5 mm. Colour as in male. Subgenital plate large, with a low incision in the middle. Sternum 9 with two darker areas laterally (Fig. 63). The dark areas are less distinct than in *Perlodes dispar.*

Nymphs. Body length 13-25 mm. Colour dark brown. Head with a lighter marking between the ocelli (Fig. 87). Tergites 7-10 with short stout bristles. Pronotum with a fringe of small bristles. Lacinia gradually pointed (Fig. 81).

Variability. Unknown. The stability of the characters used for identification may therefore be questioned.

Distribution. *P. microcephala* occurs only in Denmark, where the species is much more common than *P. dispar,* and is widely distributed. The distributional range in Europe (Fig. 20).

Biology. The nymph is predatory and occurs in rivers and large streams. Little is known about the length of the life cycle, but it probably lasts for one year.

Subfamily Isoperlinae

Males of this subfamily have a bent, finger-shaped paraproct and sclerotised teeth on penis. A well developed ventral lobe on the posterior part of sternum 8. Wings with one, seldom two cross-veins between C and R. Vein R2+3 often unforked; in Perlodinae this vein is forked. Females possess a well developed subgenital plate on sternum 8; it covers most of sternum 9.

The subfamily consists of 5 genera, only one of which is represented in Fennoscandia and Denmark.

Genus *Isoperla* Banks, 1906

Isoperla Banks, 1906, Ent. News, 17: 175.

Type species: *Sialis bilineata* Say, 1823 (des. Banks, 1906).

Medium-sized species of varying colours. Females always full-winged, males usually full-winged. Wings light brown, green or grey, veins darker and simple. Head usually broader than pronotum. Cerci and antennae long and slender. *Isoperla* species are in the female sex separated by the shape of the subgenital plate, and in the male by differences in the penial armature of the copulatory organ. The nymphs have characteristic dark-coloured bristles scattered all over the body (Fig. 91). Specific identification can be difficult on both adults and nymphs. A marked variation may occur in nearly all the taxonomical characters employed, including the ventral lobe on the posterior part of sternite 8, which has been used by some authors for identification. In the nymphs the colour patches on the head show some variation, but the largest variation occurs in the bristle arrangements on the inner margin of the lacinia (see figs 83-85). The lacinial characters used by some other authors have therefore been omitted in this description and are not used in the key.

42 species occur in Europe, three of them are found in Fennoscandia and Denmark.

7. *Isoperla difformis* (Klapálek, 1909)

Figs 68, 71, 83, 92, 97.

Chloroperla difformis Klapálek, 1909, *in* Brauer: Süsswasserfauna Deutschlands, 8: 51.

Male. Body length 6.6-8.7 mm, wing length 1.7-2.8 mm (micropterous). This small and slender stonefly has a shiny dark brown to black body. Abdominal sternum 8 with a well developed ventral lobe. Abdominal apex in ventral view (Fig. 68), and the copulatory organ possessing teeth. The penial armature is not visible in a ventral view.

Female. Body length 7.8-13.5 mm, wing length 9.0-13.1 mm. Body dark brown to black, stout and broad; usually our largest *Isoperla* female. Subgenital plate large, covering most of sternum 8 (Fig. 71).

Nymphs. First instar nymphs 0.8-1.0 mm. The fully grown male nymph 8-11 mm, female nymph 10-16 mm. Head dorsally light coloured anteriorly and dark in the posterior half (Fig. 92). Maxilla (Fig. 83). Body dark grey-green coloured, lighter ventrally. Cerci with both long and short bristles anteriorly on each segment. Long intermediate bristles present on the surface of the cercal segments (Fig. 97).

Variability. Some marked variation occurs in the structure of the genital appendages of the males, but the greatest variation is seen in the structure of the female subgenital plate (Lillehammer, 1974a). The wing venation of the female can vary markedly, and also the taxonomical characters of the nymphs undergo some variation.

Distribution. *I. difformis* occurs in all parts of Fennoscandia. In Denmark it occurs over most parts of Jutland, but not on the islands. The species occurs in northern Germany and central Europe, but not in Great Britain.

Biology. The nymph is generally predatory, though at times omnivorous. The species occurs mainly in streams in which a fine-grade substratum is present beneath the stones in the stream bed. The life cycle takes one year. Egg development is rapid at high temperatures (Saltveit & Lillehammer, 1984). Emergence in May-July, depending on altitude and latitude.

8. ***Isoperla grammatica*** (Poda, 1761)
Figs 69, 72, 84, 94, 98.

Phryganea grammatica Poda, 1761, Ins. Mus. Graecensis: 99.
Perla virens Zetterstedt, 1840, Ins. Lapp.: 1059.
Chloroperla Strandi Kempny, 1900, Verh. zool.-bot. Ges. Wien, 50: 93.
Chloroperla virens var. *annulata* Bengtsson, 1933, Lunds Univ. Årsskr. (2), 29: 26.

Male. Body length 8-14 mm, wing length 7.8-12.1 mm. Body slender, antennae and cerci long and slender. Colour of fresh specimens green to yellow-green. Head and thorax darker pigmented, head with light patches. Wings yellow-green. Sternum 8 with a well developed ventral lobe. The penial armature is visible ventrally (Fig. 69).

Female. Body length 8-12 mm, wing length 9.4-13.3 mm. Body colour same as in male. Body more slender than in *I. difformis*. Subgenital plate triangular in shape and rounded posteriorly, smaller than in *I. difformis* (Fig. 72).

Nymphs. First instar nymph 0.8-1.0 mm, fully grown nymph 9-15 mm. Head with dark markings that do not extend to the anterior margin; lighter patches around the eyes and on the middle of the head (Fig. 94). Maxilla (Fig. 84). Cercal segments 7-10 with only short bristles, a few additional surface bristles present on the segments; segments 17-19 with a marginal fringe of long bristles (Fig. 98).

Variability. Some variation occurs in the shape of the male genital appendages. The greatest variation is seen in the shape of the female subgenital plate (Lillehammer, 1974a). Only a small variation is seen in the wing venation. Marked morphological variations are also found in the nymphs.

Distribution. *I. grammatica* occurs in all parts of Fennoscandia and Denmark. The species is distributed over most of Europe, including Great Britain, Spain and Italy.

Biology. The nymph is mainly predatory, though at times omnivorous. *I. grammatica* inhabits both small and large streams, and rivers, but has not been taken in lakes. The species is supposed to have a one-year life cycle, adults emerging in May-July. The eggs, at least those of some populations, need a water temperature of 8°-12°C before the incubation starts (Saltveit & Lillehammer, 1984). Nymphs often have an unsyncronised growth, producing adults over a long span of time.

9. ***Isoperla obscura*** (Zetterstedt, 1840)
Figs 2, 3-8, 70, 73, 85, 91, 93, 99.

Perla obscura Zetterstedt, 1840, Ins. Lapp.: 1058.
Perla griseipennis Pictet, 1841, Perlides: 299.
Chloroperla limbata Bengtsson, 1933, Lunds Univ. Årsskr. (2), 29: 26.

Male. Body length 7.1-10.5 mm, wing length 7.7-9.6 mm, always full-winged. Body grey-green and brown-grey and usually dark-pigmented. Head with yellow patches. Wings smokey grey with darker veins. Sternum 8 with a well-developed ventral lobe. Penial armature invisible ventrally (Fig. 70).

Female. Body length 8.0-12.5 mm, wing length 9.0-12.8 mm. Body colour same as in male. Subgenital plate more or less triangular in shape and pointed posteriorly (Fig. 73).

Nymphs. First instar 0.8-1.0 mm, fully grown nymphs 9-13 mm. Body yellow-green. Head darker pigmented anteriorly, with only a dark area in the middle of the head; also dark pigmented on the anterior part of the labrum (Fig. 93). Maxilla (Fig. 85). Cercal segments 7-10 with only short bristles, but some additional intermediate bristles on the segmental surface; segments 17-19 with a marginal fringe of long bristles (Fig. 99).

Variability. Variation occurs both in the male genital appendages and in the subgenital plate of the female (Lillehammer, 1974a). Only slight variation occurs in the wing venation. A marked variation is seen in nymphal morphology.

Distribution. *Isoperla obscura* occurs in Finland, Norway and Sweden, but not in Denmark. Most abundant at high altitudes and in more northern latitudes. Can also be fairly common in larger rivers in the south. The species is recorded from northern Germany and in central Europe. Old records exist also from Great Britain. The species seems to be the dominating stonefly predator in northenmost Fennoscandia.

Biology. The nymph is mainly predatory, though at times omnivorous. *I. obscura* occurs both in small streams and rivers, and also in mountain lakes. In the southern parts mainly recorded from large rivers. The species has a two-year life cycle, with an egg diapause lasting 8-10 months during the first winter. At high altitudes the species seems to require two summers for the nymphal growth. Emergence takes place in July-

August (Lillehammer, 1985c). The life cycle of lowland populations has not yet been properly studied.

Family Perlidae

Large, grey-brown or black stoneflies. Wings smoky grey with dark-pigmented veins. Females usually much larger than males. Nymphs large, dark grey-brown. Femur and tibia with a dense fringe of long bristles posteriorly. Zwick (1973) based the monophyly of this family on the following synapomorphic characters: the small number (6) of abdominal ganglia, and the pronounced glossa and the broad mentum of the nymph.

The family consists of sixteen genera. In Fennoscandia and Denmark only one genus and species occurs.

Genus *Dinocras* Klapálek, 1907

Dinocras Klapálek, 1907b, Rozpr. české Acad. Cís. Fr. Jos., II, 16: 4.
Type species: *Perla cephalotes* Curtis, 1827 (des. Klapálek, 1907).

Body dark brown to black, with yellow spots on the head. Head broader than pronotum. Male short-winged, female full-winged. Cerci long and stout. A typical feature of all *Dinocras* species is that the cross-vein r-m in the forewing arises from Rs before this divides into R2+3 and R4+5, and the presence of several cross-veins between C and Sc (Fig. 42). Hind wing with one or more cross-veins between M and Cu. Nymphs (Fig. 52) large, stout, brown or dark grey, and dorso-ventrally flattened. Filamentous gills present on thorax and on the subanal plates. Cerci large with stout bristles on the segments. *Dinocras* is a European genus with 3 species.

10. ***Dinocras cephalotes*** (Curtis, 1827)
Figs 11, 14, 22, 42, 52, 100, 101.

Perla cephalotes Curtis, 1827, Brit. Ent., 4: 455.
Perla baetica Rambur, 1845, Hist nat. Ins., Névr.: 455.
Perla elegantula Klapálek, 1905b, Čas. české Spol. ent., 2: 29.

Male. Body length 13.5-19 mm, wing length 8.5-13 mm (brachypterous). Tergum 10 bilobed (Fig. 100). Body grey to black, with yellow spots on the head. Thorax dark-pigmented, wings smoky grey with darker veins.

Female. Body length 19.5-29 mm, wing length 19-29 mm. Subgenital plate triangular in form and covering most of sternum 8 and nearly half of sternum 9 (Fig. 101). Body colour of head and thorax same as in male, abdomen yellow-brown. *Dinocras* females are the largest stoneflies occurring in Fennoscandia and Denmark, and they are always full-winged.

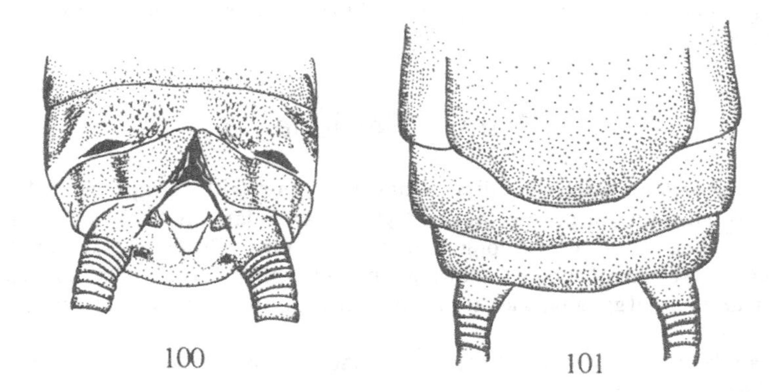

Figs 100, 101. *Dinocras cephalotes* (Curt.). - 100: male abdominal apex in dorsal view; 101: female abdominal apex in ventral view.

Nymphs. First instar 1.0-1.4 mm. Fully grown male nymphs 12.5-17 mm, female nymphs 20-31 mm. Nymphs dark brown to grey-brown on the dorsal side and lighter ventrally. The nymph is heavily sclerotised, and the body is stout and broad with long and stout cerci and antennae. Filamentous gills present on all thoracic segments (Fig. 52). The largest plecopteran nymph found in Fennoscandia and Denmark.

Variability. The form of the genital appendages and the wing venation pattern are fairly constant, but body length and wing length of both species varies considerably.

Distribution. *D. cephalotes* occurs in Denmark, Sweden and Norway, but not in Finland (Fig. 22). The species is a south-western immigrant to Fennoscandia, and has a scattered distribution. It is most common in the lowland, but has been recorded in the subalpine vegetation zone at about 1000 m a. s. l. In Denmark the species has only been recorded from one locality in western Jutland.

Biology. The nymph is predatory, feeding largely on larvae of Chironomidae and nymphs of Ephemeroptera. *D. cephalotes* has a three- to four-year life cycle, adult emerging in May-July. The eggs require a water temperature of about 12°C before development starts (Lillehammer, 1987b). *D. cephalotes* occurs both in small streams and in large rivers, but not in lakes.

Family Chloroperlidae

Medium-sized to small stoneflies, light green to yellow in colour. Both sexes full-winged, anal area of hindwing strongly reduced (Fig. 9). The family is characterised by two synapomorphic characters: the reduced terminal segment of the maxillary palp (Fig. 47), and reduction of the caecal sacs of the gut.

The family is divided into two subfamilies: Paraperlinae and Chloroperlinae, only the latter being represented in Europe. 14 genera of Chloroperlinae are recognised; three of these occur in Fennoscandia.

Key to species of Chloroperlidae

Adults

1 Mesonotum with a U-shaped dark-pigmented patch (Fig. 102). Antenna serrate. Male: tooth of epiproct large (Fig. 104). Female: subgenital plate large and rectangular (Fig. 107) .... 11. *Isoptena serricornis* (Pictet)

\- Mesonotum with a W-shaped dark-pigmented patch (Fig. 103). Antenna filiform 2

2(1) Male: tooth of epiproct small and weakly pigmented (Fig. 106). Female: subgenital plate small and well pigmented (Fig. 109) .... 13. *Xanthoperla apicalis* Newman

\- Male: tooth of epiproct large and heavily pigmented (Fig. 105). Female: subgenital plate large and weakly pigmented (Fig. 108)......................... 12. *Siphonoperla burmeisteri* (Pictet)

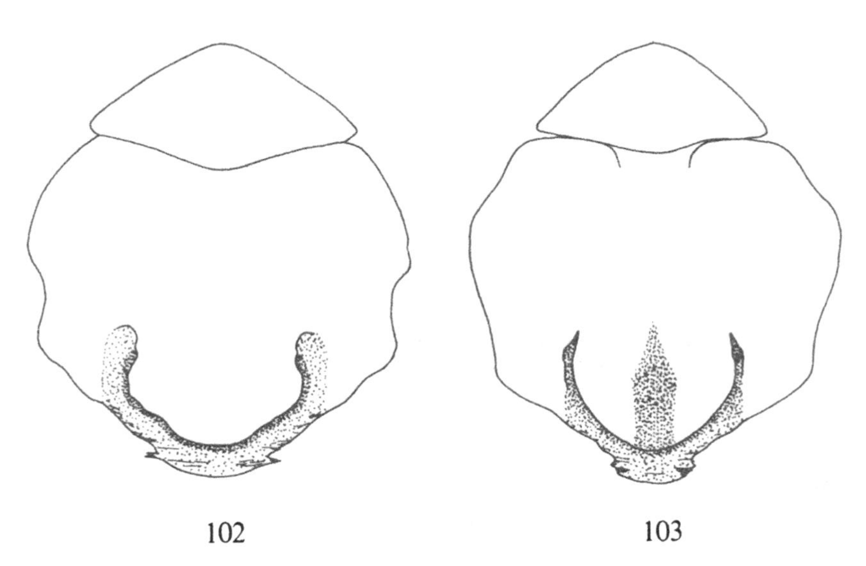

Figs 102, 103. Mesonotal pattern in Chloroperlidae. - 102: *Isoptena;* 103: *Siphonoperla* and *Xanthoperla.*

Figs 104-106. Male abdominal apex in Chloroperlidae, lateral view. - 104: *Isoptena serricornis* (Pict.); 105: *Siphonoperla burmeisteri* (Pict.); 106: *Xanthoperla apicalis* Newm.

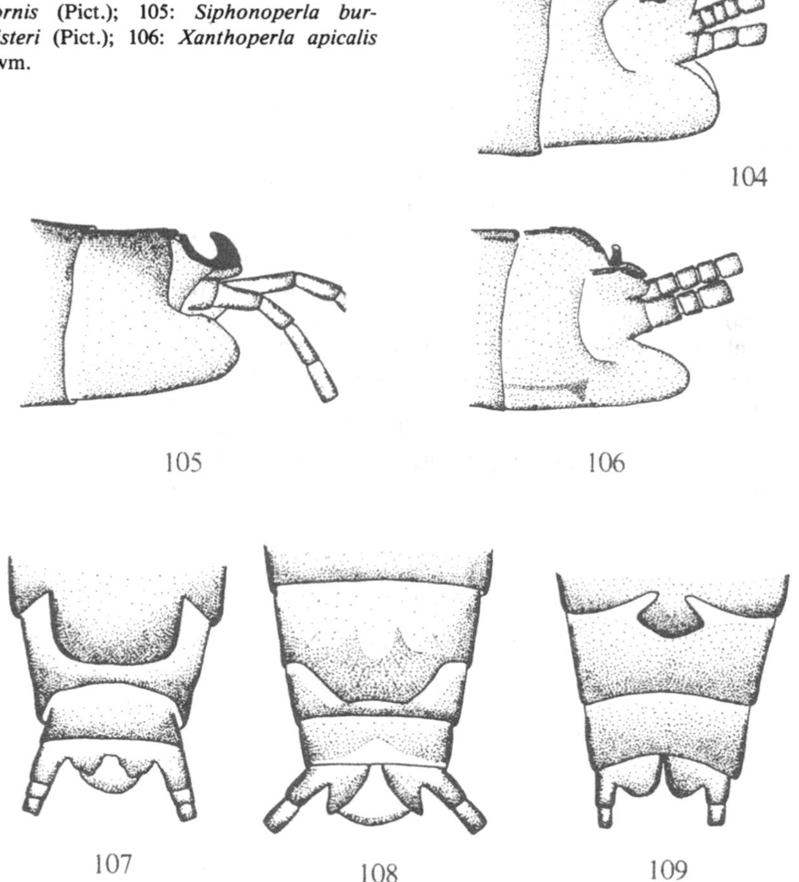

Figs 107-109. Female abdominal apex in Chloroperlidae, ventral view. - 107: *Isoptena serricornis* (Pict.); 108: *Siphonoperla burmeisteri* (Pict.); 109: *Xanthoperla apicalis* Newm.

Nymphs

1 Antennae reduced, first segment with long bristles (Fig. 110). Whole body including pronotum (Fig. 112), as well as femur and tibia (Fig. 115) with a dense covering of long bristles .... 11. *Isoptena serricornis* (Pictet)

\- Antennae normal (Fig. 111). Scattered bristles present over whole body 2

2(1) Body-shape short and stout. Head wider than long. Pronotal bristles long and pointed (Fig. 113). Tibia with a dense fringe of long bristles (Fig. 116) .... 12. *Siphonoperla burmeisteri* (Pictet)
- Body-shape elongate and slender. Head longer than wide. Pronotal bristles long and blunt-tipped (Fig. 114). Tibia with a sparse fringe of long bristles (Fig. 117) .. 13. *Xanthoperla apicalis* Newman

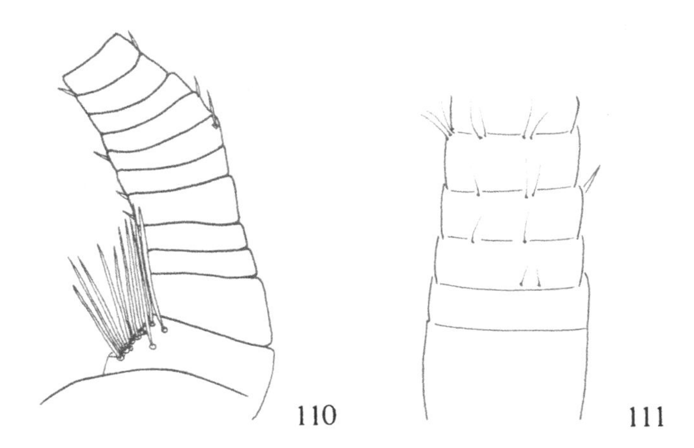

Figs 110, 111. First antennal segments of nymphs of 110: *Isoptena serricornis* (Pict.) and 111: *Siphonoperla burmeisteri* (Pict.).

Genus *Isoptena* Enderlein, 1909

Isoptena Enderlein, 1909, Zool. Anz., 34: 414.
Type species: *Perla serricornis* Pictet, 1841 (mon.).

Slender yellow-brown stoneflies. Antennae serrate, the first 8-10 segments lighter than the following. Pronotum with a lighter band medially. Mesonotum vith a U-shaped dark-pigmented patch (Fig. 102). Forewing with 3 anal veins. Nymphs yellow to light brown with long and slender bristles on the head, thorax and legs.

Only one West Palaearctic species.

11. ***Isoptena serricornis*** (Pictet, 1841)
Figs 27, 102, 104, 107, 110, 112, 115.

Perla (Isopteryx) serricornis Pictet, 1841, Perlides: 303.

Male. Body length 7-8 mm, wing length 7-9 mm. Body colour brown, wings light brown. Abdominal terga 2-8 with a dark mid-band. A characteristic hook present on tergum 10 (Fig. 104).

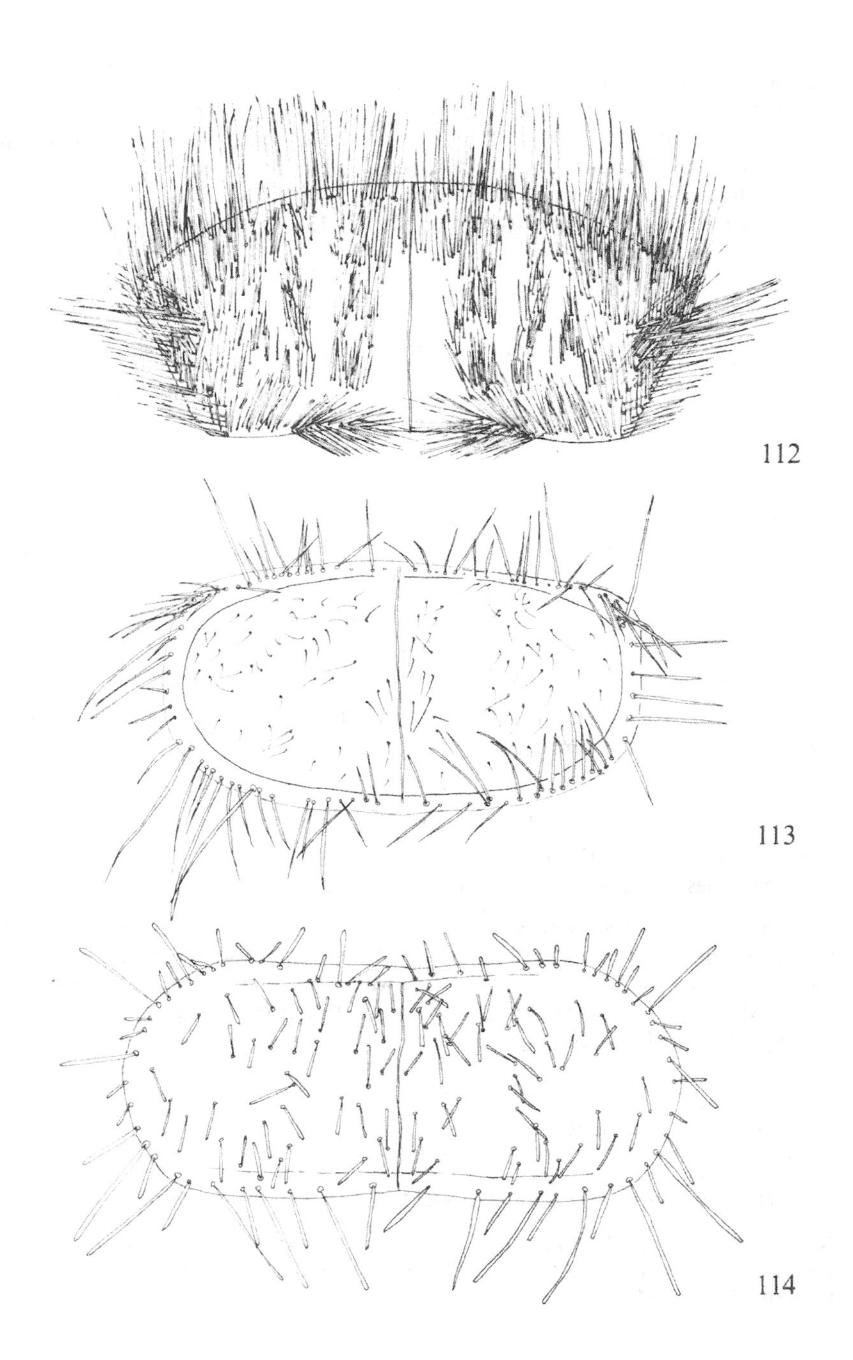
112
113
114

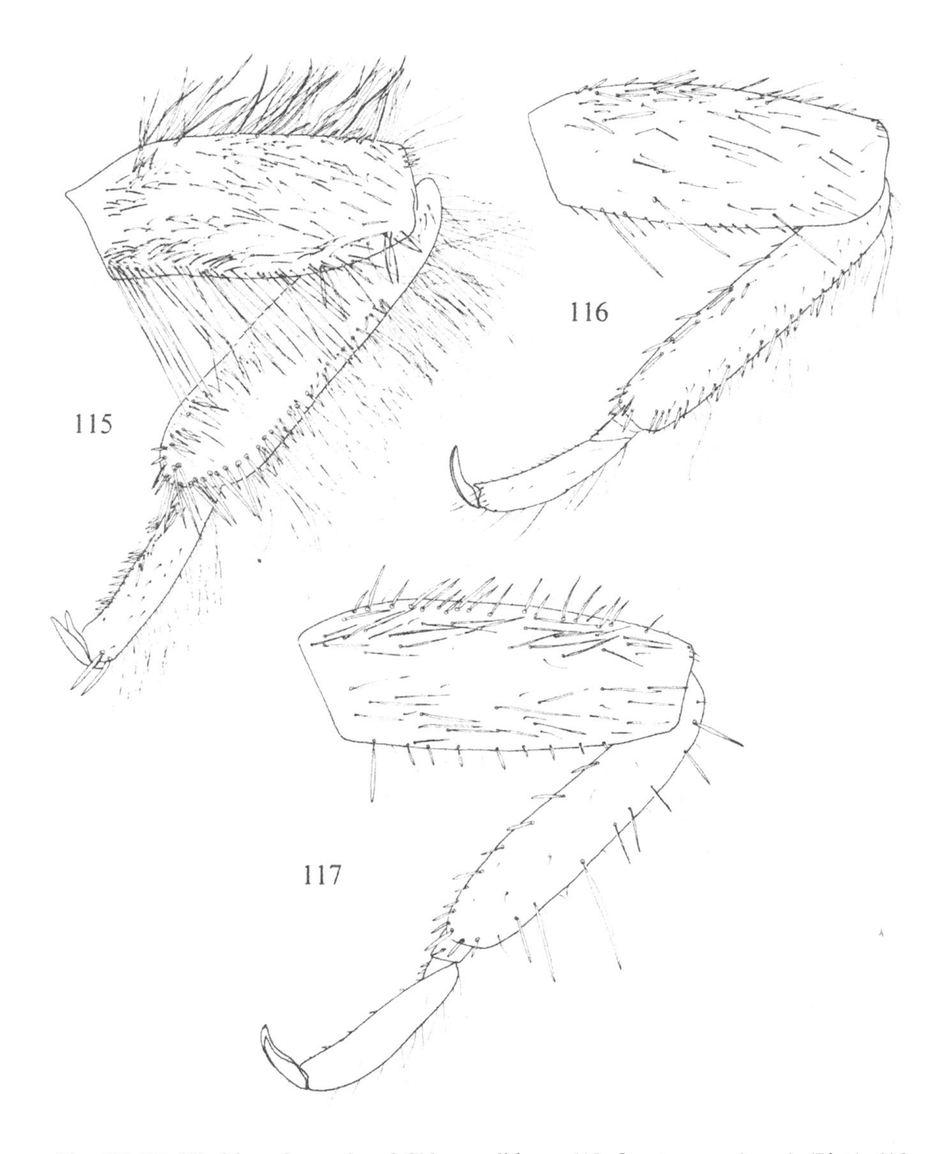

Figs 115-117. Hind leg of nymphs of Chloroperlidae. - 115: *Isoptena serricornis* (Pict.); 116: *Siphonoperla burmeisteri* (Pict.); 117: *Xanthoperla apicalis* Newm.

Figs 112-114. Pronotum of nymphs of Chloroperlidae. - 112: *Isoptena serricornis* (Pict.); 113: *Siphonoperla burmeisteri* (Pict.); 114: *Xanthoperla apicalis* Newm.

Female. Body length 9-11 mm, wing length 9-12 mm. Body colour same as in male. Subgenital plate large and rectangular (Fig. 107).

Nymphs. Body yellow. Segments 2-4 of maxillary palps flattened. Tufts of hairs present around the eyes. Antennae reduced in size and with a lateral row of bristles on first segment (Fig. 110). Pronotum (Fig. 112) and legs (Fig. 115) densely covered by long bristles.

Variability. The material studied is too sparse for any conclusion to be drawn.

Distribution. Denmark: *I. serricornis* occurs in central parts of Jutland but has not been recorded from Funen, Zealand, or any of the other islands. In Finland and Sweden the species has mainly been recorded from some few northern districts. It has not yet been recorded from Norway. The distribution pattern indicates that the species has entered Denmark from the south-east and Fennoscandia from the north-east. The European distribution (Fig. 27).

Biology. The species is mainly recorded from rivers with a sandy bed. The nymph is supposed to be a burrowing animal. *I. serricornis* probably has a one-year life cycle, adults emerging in May-July.

Genus *Siphonoperla* Zwick, 1967

Siphonoperla Zwick, 1967, Mitt. schweiz. ent. Ges., 40: 10.

Type species: *Perla torrentium* Pictet, 1841 (orig. des.).

Small yellow-green stoneflies. Meso- and metanotum with a w-shaped dark-pigmented patch (Fig. 103). The genus is characterised by the shape of the male genitalia and the female subgenital plate. Penis is covered by a sack up to the base of the penial shaft. The female subgenital plate is large, broader than long. Nymphs yellow-brown, with broad head and short stout cerci.

The genus is Vest Palaearctic and has 11 species; only one occurs in Fennoscandia.

12. *Siphonoperla burmeisteri* (Pictet, 1841)

Figs 9, 103, 105, 108, 111, 113, 116.

Perla (Isopteryx) burmeisteri Pictet, 1841, Perlides: 311.

Leptomeres rufeola Rambur, 1842, Hist. nat. Inst., Névr.: 457.

Male. Body length 5.6-7.2 mm, wing length 6.3-7.2 mm. Head broader than long, with marked dark spots between the ocelli. The basal antennal segments dark brown pigmented. Body slender, yellow-green, wings yellow with weakly pigmented veins. The genitalic hook on segment 10 is large and darkly pigmented (Fig. 105). Penis with two titillators and groups of penial teeth.

Female. Body length 5.0-8.0 mm, wing length 6.5-8.4 mm. Body colour and wing as in male. Subgenital plate large, lightly pigmented, covering most of segment 9 (Fig. 108).

Nymph. First instar 0.80-0.96 mm. The fully-grown nymph 6-9 mm. Body yellow-brown, stout, with strong and characteristically short cerci. Head broader than long, antenna (Fig. 111) normal. Pronotal bristles long and pointed, forming a more or less continuous fringe (Fig. 113). Tibia with a dense fringe of long bristles (Fig. 116). Cerci short and stout with broad segments.

Variability. Some variation occurs in body length, venation and in wing length. There occur some forking in both R2+3 and R4+5. Some specimens show a tendency to be short-winged. Some variation also occurs in the form of the female subgenital plate. There occur specimens which have a rounded apex and others with an incision. However, the form shown in Fig. 108 is the most common.

Distribution. *S. burmeisteri* occurs in Sweden, Norway and Finland; in Denmark in parts of Jutland, but not on the islands. The species also occurs in the eastern parts of central Europe, and in Asia Minor. In Fennoscandia the species occurs both in the most continental areas as well as in coastal areas in the west.

Biology. The species occurs in both large and small streams and also in the exposed zone of lakes. It has not been recorded from sites above the sub-alpine zone. *S. burmeisteri* has a one-year life cycle, and adults emerge in June-July. The eggs require a water temperature of 4-8°C before development starts (Lillehammer, 1987c).

Genus *Xanthoperla* Zwick, 1967

Xanthoperla Zwick, 1967, Mitt. schweiz. ent. Ges., 40: 8.
Type species: *Chloroperla apicalis* Newman, 1836 (orig. des.).

Small, yellow-green stoneflies. Mesonotum with a w-shaped dark-pigmented patch (Fig. 103). The genus is characterised by the shape of the male genitalia and the shape of the female subgenital plate. Penis is completely covered by a sack. Female subgenital plate small and short. Nymphs yellow, head longer than wide, without dark spots between the ocelli; antennae and cerci long and slender.

13. *Xanthoperla apicalis* (Newman, 1836)
Figs 103, 106, 109, 114, 117.

Chloroperla apicalis Newman, 1836, Ent. Mag., 3: 501.
Chloroperla pallida Stephens, 1836, Illus. Brit. Ent., 6: 139.
Leptomeres pallidella Rambur, 1842, Hist. nat. Inst., Névr.: 458.
Isopteryx hamulata Morton, 1930, Entomologist's mon. Mag., 66: 78.

Male. Body length 5.5-7.0 mm, wing length 6.0-6.9 mm. Head longer than wide, without dark spots between the ocelli. Antennae long and slender. Meso- and metanotum with dark w-shaped pigmentation. Body light yellow, wings yellow with lightly pigmented veins. The veins are often so lightly pigmented that they are difficult to observe in old alcohol preserved specimens. Genital hook on segment 10 is small and brown in colour (Fig. 106).

Female. Body length 6.7-8.1 mm, wing length 6.8-7.9 mm. Body colour same as in the male. Subgenital plate small and darkly pigmented (Fig. 109), covering only a small part of segment 9.

Nymphs. Small and slender, head longer than broad, whole body yellow to light brown. Head with long and slender antennae, pale area present between ocelli. Pronotal bristles long and blunt-tipped (Fig. 114). Tibia with a thin fringe of long bristles (Fig. 117). Cerci long and slender. The nymph has recently been described (Brittain, 1983b). The nymphs can easily be separated from *S. burmeisteri* by the long and slender, slightly pigmented body with the elongated head and the long and slender cerci.

Variability. Has not been studied.

Distribution. In Finland and Sweden, *X. apicalis* occurs only in the northern parts, while in Norway it also occurs in the south. The species is absent from Denmark. *X. apicalis* is supposed to be a north-eastern immigrant to Fennoscandia. The species occurs in northern Germany, France, Spain, Italy and eastern Europe.

Biology. The species occurs mainly in rivers, in southern Norway only in large rivers such as Glomma and Namsen. The species is supposed to have a one-year life cycle, and adult emergence takes place in June and July.

Family Taeniopterygidae

Small to medium-sized species. Both sexes long-winged. Sternum 9 of male enlarged, forming a well developed subgenital plate. Female subgenital plate minute or absent, but sternite 9 enlarged and forming a postgenital plate. Female genital opening situated in middle of sternum 8 (Fig. 124).

Zwick (1973) mentioned the following apomorphies of the family: all three tarsal segments approximately equally long (Fig. 35); the second tarsal segment straight; and the first cercal segment of the male more or less (often very) swollen and occasionally with hooks.

Brinck (1952) separated *Taeniopteryx* from *Brachyptera* on the following characters: in *Brachyptera* the anterior cubital vein sends 3-4 branches to the wing edge while *Taeniopteryx* has at the most 2 branches. Illies (1955) and Zhiltsova (1964) stated that there are usually 3-4, seldom 2, branches in *Brachyptera*. Brinck (1952) and Hynes (1967) separated the genus *Rhabdiopteryx* from *Taeniopteryx* on the following characters: *Rhabdiopteryx* has 1-3 cross-veins between C and Sc, *Taeniopteryx* has none. Illies (1955) gave 2-4 or more cross-veins for *Rhabdiopteryx* and none for *Taeniopteryx*. Lillehammer (1974a) showed that some of these characters overlapped to such a degree that they are of little value for identification.

The family is divided into two subfamilies, the Taeniopteryginae with one genus, *Taeniopteryx*, and the Brachypterinae with 8 genera, two of which occur in Fennoscandia and Denmark.

Key to species of Taeniopterygidae

Adults

1 Each coxa with a scar on the inner side (Fig. 118). Male cerci one-segmented, female cerci with 8 or 9 segments. Male genitalia (Fig. 120). Female genitalia (Fig. 124) .... 14. *Taeniopteryx nebulosa* (Linnaeus)

\- Coxa without a scar on the inner side 2

2(1) Forewing with 2 or more cross-veins between C and Sc (Fig. 119). Male cerci with 5 segments, basal segment with a dorsal knob. Female cerci with 4-5 visible segments. Male genitalia (Fig. 123). Female genitalia (Fig. 127) .... 17. *Rhabdiopteryx acuminata* Klapálek

\- Forewing without cross-veins between C and Sc. Male cerci with two segments, basal segment much larger than second segment. Female cerci with one visible segment 3

3(2) Male subgenital plate narrow at apex (Fig. 122). Basal segment of male cerci rounded. Apex of epiproct elongate. Female postgenital plate large and pointed at apex (Fig. 126) .... 16. *Brachyptera risi* (Morton)

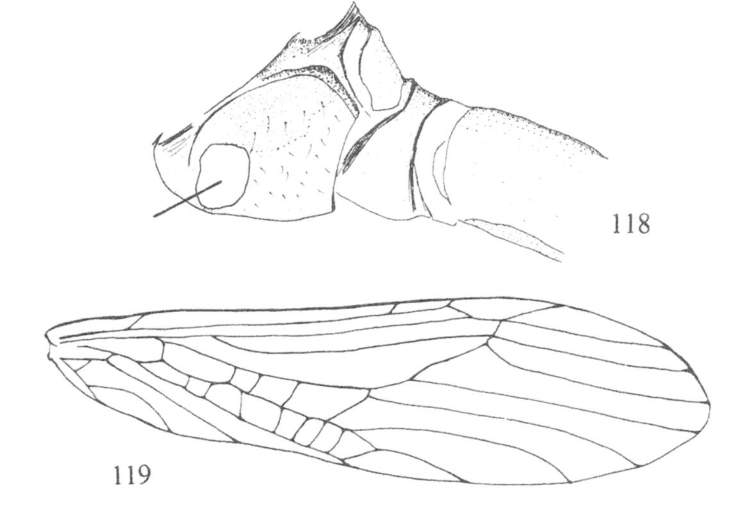

Fig. 118. Coxa of *Taeniopteryx nebulosa* (L.).
Fig. 119. Right forewing of *Rhabdiopteryx acuminata* Klap.

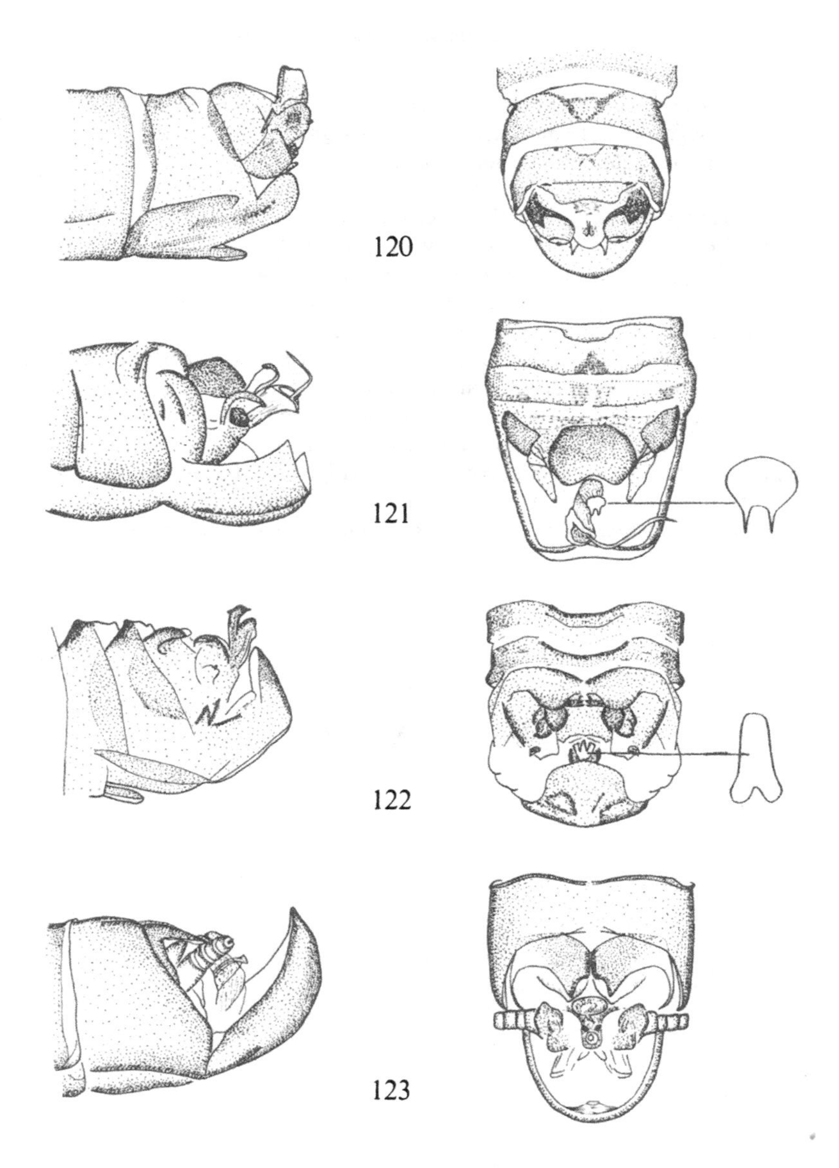

Figs 120-123. Male abdominal apex of Taenioterygidae, left row in lateral view, right row in dorsal view. - 120: *Taeniopteryx nebulosa* (L.); 121: *Brachyptera braueri* (Klap.); 122: *B. risi* (Mort.); 123: *Rhabdiopteryx acuminata* Klap.

– Male subgenital plate broad at apex (Fig. 121). Basal segment of male cerci elongate. Apex of epiproct rounded. Female postgenital plate large and rounded at apex (Fig. 125) .... 15. *Brachyptera braueri* (Klapálek)

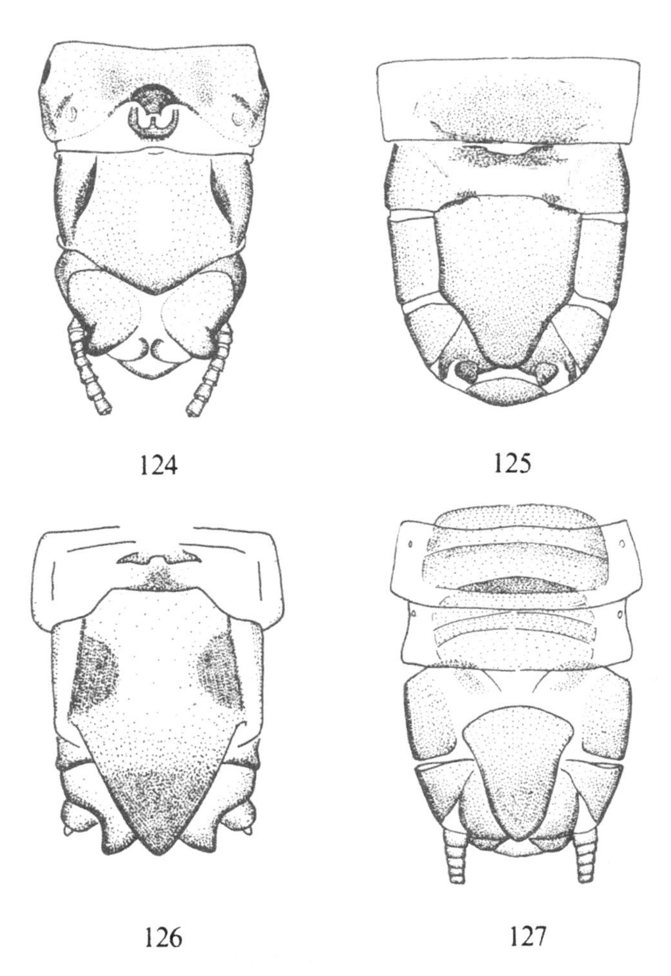

Figs 124-127. Female abdominal apex of Taeniopterygidae, ventral view. – 124: *Taeniopteryx nebulosa* (L.); 125: *Brachyptera braueri* (Klap.); 126: *B. risi* (Mort.); 127: *Rhabdiopteryx acuminata* Klap.

Nymphs

1 Segmented finger-like gills present on each coxa. Abdominal terga with knob-like processes (Fig. 128) .... 14. *Taeniopteryx nebulosa* (Linnaeus)

\- Coxae without gills. Abdominal terga without processes 2

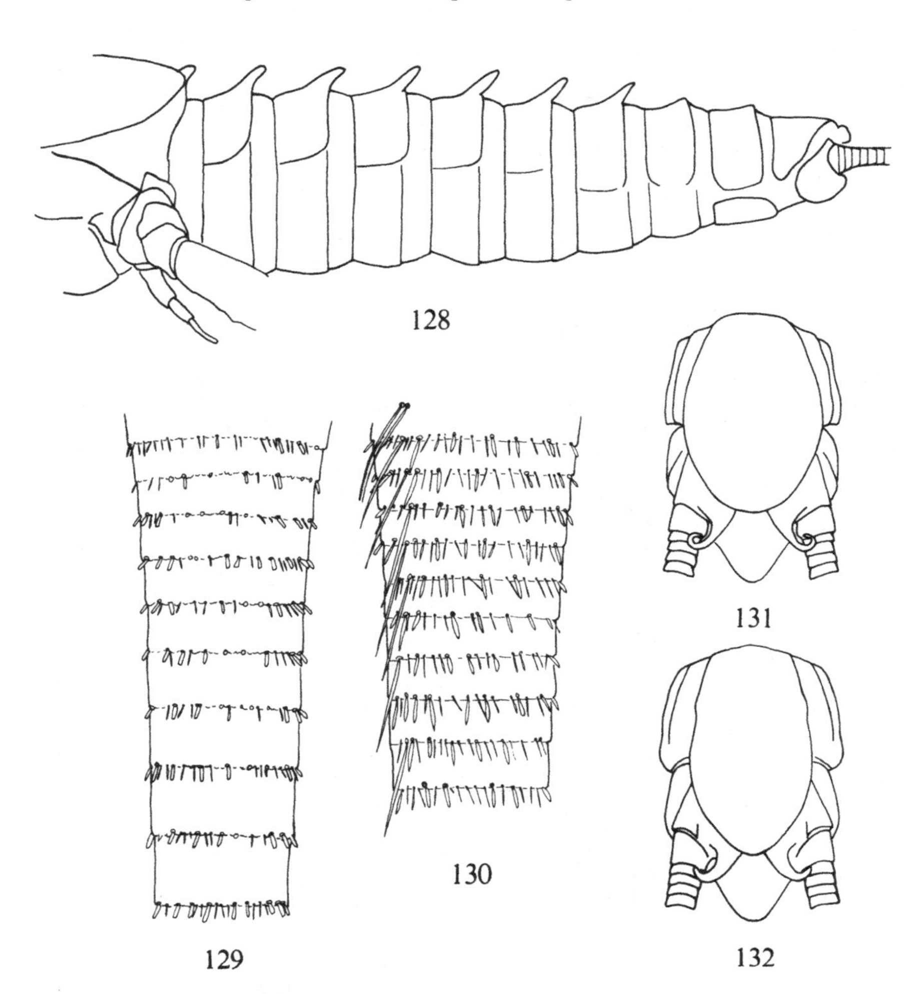

Figs 128-132. Nymphs of Taeniopterygidae. - 128: metathorax and abdomen in lateral view of *Taeniopteryx nebulosa* (L.); 129: cercal segments of ***Rhabdiopteryx acuminata*** Klap.; 130: same of ***Brachyptera risi*** (Mort.); 131: abdominal apex of ***Brachyptera braueri*** (Klap.); 132: same of ***B. risi*** (Mort.).

2(1) Cerci with long hairs on upper side (Fig. 130). Head with a U-shaped epicranial suture (Fig. 133) 3
- Cerci without long hairs on upper side (Fig. 129). Head with a V-shaped epicranial suture (Fig. 134) .... 17. *Rhabdiopteryx acuminata* Klapálek
3(2) Apex of male paraproct curled (Fig. 131). 15. *Brachyptera braueri* (Klapálek)
- Apex of male paraproct bent (Fig. 132) .... 16. *Brachyptera risi* (Morton)

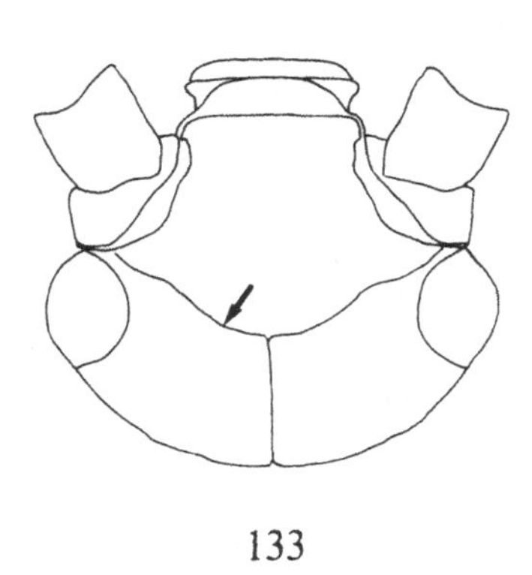

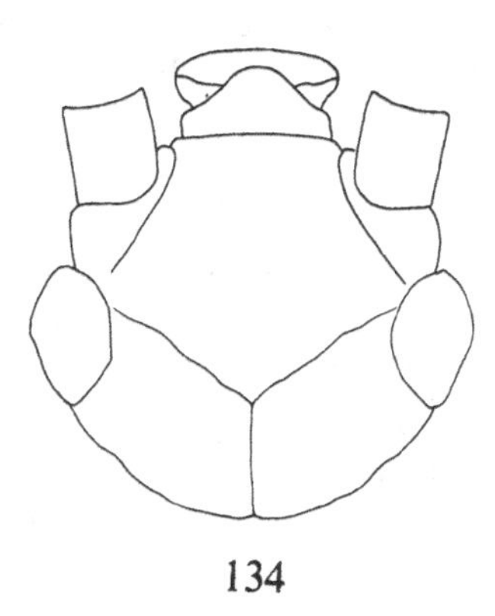

Figs 133, 134. Head of nymphs in dorsal view of 133: *Brachyptera* sp., and 134: *Rhabdiopteryx acuminata* Klap.

SUBFAMILY TAENIOPTERYGINAE

The following synapomorphic characters are mentioned by Zwick (1973): male cerci one-segmented; the receptaculum seminis consists of a bladder-like section and a tube-like section; abdomen with well developed longitudinal tracheae ventrally; and nymph with coxal gills (Fig. 128).

Genus *Taeniopteryx* Pictet, 1841

Nemoura (Taeniopteryx) Pictet, 1841, Perlides: 345.
Type species: *Phryganea nebulosa* Linnaeus, 1758 (orig. des.).

Medium-sized, dark-pigmented species with smoky grey wings. Head and pronotum of about same width. Inner side of coxa with scar from the nymphal gill. Male subgenital plate short (Fig. 120). Male with one-segmented cerci, female with several segments in the cerci. Nymphs with coxal gills and abdominal terga with knobs (Fig. 128).

Six species occur in Europe, only one in Fennoscandia and Denmark,

14. ***Taeniopteryx nebulosa*** (Linnaeus, 1758)
Figs 118, 120, 124, 128.

Phryganea nebulosa Linnaeus, 1758, Syst. Nat., 10. Ed., 1: 549.
Nemoura nigripes Zetterstedt, 1840, Ins. Lapp.: 1056.

Male. Body length 6.9-10 mm, wing length 9-13 mm. Head and thorax dark brown, abdomen rufous brown. Antennae and legs very long. Male subgenital plate about as long as wide, at base with a vesicle (Fig. 120). Cerci with only one segment.

Female. Body length 6.7-15 mm, wing length 10.2-16.2 mm. Colour same as in male. Cerci with 8 or 9 segments. Subgenital plate small, postgenital plate large (Fig. 124).

Nymphs. First instar 0.7-0.8 mm, fully grown nymphs 8-15 mm. Head and prothorax of about same width. Legs long and slender, coxae with segmented finger-like gills (Fig. 128). Abdominal terga 1-7 with knobs (Fig. 128). Body dark brown to rufous brown.

Variability. The form of the genital appendages is fairly constant, but the wing venation shows marked variation. This applies especially to those veins frequently used to separate this genus from other genera of Taeniopterygidae.

Distribution. *T. nebulosa* occurs in most parts of Fennoscandia. In Denmark it occurs in Jutland and on the island of Funen but not on Zealand. The species is distributed over most of Europe, including Great Britain, Spain, Italy and eastern Europe.

Biology. The nymphs eat leaf fragments and detritus and occur in small streams as well as in large rivers. In northern Fennoscandia they are also found in lakes. The species has a one-year life cycle with a short egg incubation period. The main growth of the nymphs takes place in summer or in autumn and early winter. Emergence of adults are seen in March-July, depending on altitude and latitude.

Subfamily Brachypterinae

The following synapomorphic characters for the members of this subfamily are given by Zwick (1973): the particular shape on sternite 9 in the male, the presence of a postgenital plate (of sternum 9) in the female, the strongly developed epiproct which is provided with an incurved sclerotised sac, and the extremely complex and richly segmented asymmetrical paraprocts. The large subanal plates of the nymphs are also characteristic.

Genus *Brachyptera* Newport, 1849

Nemoura (Brachyptera) Newport, 1849, Proc. linn. Soc. Lond., 1: 389.
Type species: *Nemoura trifasciata* Pictet, 1832 (des. Klapálek, 1902).

Medium-sized, dark brown species. Wings grey-brown with darker bands. Head and pronotum of about the same width. Male sternite 9 forms a subgenital plate that covers the genital appendages ventrally (Figs 121, 122). Female genital opening free and sternite 9 forms a large postgenital plate. Cerci with at most 4 or 5 segments. Nymphs without coxal gills, abdominal terga 1-7 without knobs. Sternite 9 enlarged posteriorly. Subanal plates of male nymphs pointed posteriorly. Antennae, legs and cerci long and slender.

Fourteen species of *Brachyptera* occur in Europe; only two of them are found in the Fennoscandian fauna.

15. *Brachyptera braueri* (Klapálek, 1900)

Figs 121, 125, 131, 133

Taeniopteryx Braueri Klapálek, 1900, Rozpr. české Akad. Cís. Fr. Jos., II, 9: 7.

Male. Body length 8.0-10.0 mm, wing length 8.9-10.1 mm. Dark brown to black, abdomen brown, wings smoky grey. Genital appendages (Fig. 121) of complex structure: apex of epiproct rounded, supra-anal lobe broadening at apex, and cerci with two segments; basal segment elongate, while second segment is small and rounded.

Female. Body length 9.0-13.0 mm, wing length 12.6-13.5 mm. Dark brown to black, abdomen lighter brown, wings smoky grey. Postgenital plate large and rounded apically (Fig. 125). Cerci one-segmented.

Nymphs. Fully grown nymph 8-10 mm. Body rufous to light brown and yellow. Head with a U-shaped epicranial suture (Fig. 133). Cerci with long bristles dorsally (as in Fig. 130). Apex of male paraproct curled (Fig. 131).

Variability. The degree of variation has not yet been studied.

Distribution. *B. braueri* occurs in southern Sweden and in a few localities in Jutland, Denmark. The species has not been recorded from Norway or Finland, but occurs in northern, central and eastern Europe; not in Great Britain.

Biology. The species is supposed to have a one-year life cycle. It occurs in rivers and larger streams. Emergence takes place in March-April.

16. *Brachyptera risi* (Morton, 1896)

Figs 21, 122, 126, 130.

Taeniopteryx risi Morton, 1896, Trans. ent. Soc. Lond., 1896: 56.
Nemoura variegata Stephens, 1836, Illus. Brit. Ent., 6: 144.

Male. Body length 6.3-11.0 mm, wing length 9.5-11.7 mm. Body dark brown or black, wings smoky grey with darker bands. The genital appendages of complex structure: apex of epiproct elongate, supra-anal lobe pointed at apex, and cerci with 2 segments; basal segment large and rounded (Fig. 122).

Female. Body length 7-13 mm, wing length 10.3-14 mm. Body dark brown to black, abdominal segments lighter in colour, wings smoky grey with darker bands. Postgenital plate very large and pointed apically (Fig. 126). Cerci one-segmented.

Nymphs. Fully grown nymphs 7-12 mm, rufous brown or green-brown. Cerci with long hairs on the upper side (Fig. 130). Head with a U-shaped epicranial suture (Fig. 133). Apex of male paraproct bent (Fig. 132).

Variability. The form of the genital appendages is fairly constant. A marked variation in the cubitus vein (Cu1) occurs.

Distribution. *B. risi* occurs over most of Sweden, Norway and Denmark. In Finland only in the north (Fig. 21). The species is supposed to be a southwestern immigrant to Fennoscandia. It occurs over most of Europe, including Great Britain.

Biology. *B. risi* occurs in streams and rivers, but not in lakes. The nymph grazes on the periphyton. Studies on oxygen consumption have shown that nymphs of *risi* have a greater oxygen requirement than nymphs of other comparable species. It is supposed to have a one-year life cycle, adults emerging in March-July.

Genus ***Rhabdiopteryx*** Klapálek, 1902

Rhabdiopteryx Klapálek, 1902, Természetr. Füz., 25: 179.

Type species: *Taeniopteryx hamulata* Klapálek, 1902 (des. Klapálek, 1905).

Medium-sized species, body dark brown and with smoky grey wings. Both sexes are long-winged. Head and pronotum of about the same width. Antennae long and slender. Male subgenital plate large, overreaching abdominal apex (Fig. 123). Both sexes have several segments in the cerci. Nymphs without coxal gills or tergal knobs on the abdominal terga. Sternum 9 of nymph enlarged.

17. ***Rhabdiopteryx acuminata*** Klapálek, 1905
    Figs 119, 123, 127, 129, 134.

Rhabdiopteryx acuminata Klapálek, 1905a, Čas. české Spol. ent., 2: 10.
Rhabdiopteryx anglica Kimmins, 1943, Proc. R. ent. Soc. Lond., (B), 12: 42.

Male. Body length 9.0 mm, wing length 10.0 mm. Head, pronotum, and meso- and metanotum dark brown. Antennae long and dark brown, third segment much longer than the rest. Legs long and of a light brown colour. Abdominal segments dark brown. Subgenital plate very large and overreaching abdominal apex (Fig. 123).

Female. Body length 10.0 mm, wing length 11.0 mm. Body colour as in male, except for the abdominal segments which are lighter brown. Postgenital plate large, narrow, and pointed apically (Fig. 127).

Nymph. Fully grown nymph 10-12 mm long. Head, antennae and thorax dark brown with lighter patches; abdominal segments brown. Femora and tibiae with a dense fringe of long fine bristles. Cercal segments with short stout bristles (Fig. 129). Head with a Y-shaped epicranial suture (Fig. 134).

Variability. Not known.

Distribution. Until now this species had only been recorded from a locality near Helsinki in Finland. It can now also be recorded from Joensuu in Kb.

Biology. The species occurs in a fast flowing stream. The substratum consists of mostly gravel with scattered stones, covered by *Fontinalis* spp. Emergence in early May.

Family Nemouridae

Small to middle-sized species, dark brown to black in colour. Wings light brown or grey, with cross-veins in the cubital area, Scl and Sc2 present in both wings. The X-pattern of the wing venation, used by several authors to separate this family from all other families is an invalid character (Lillehammer, 1974a) (Figs 135-137).

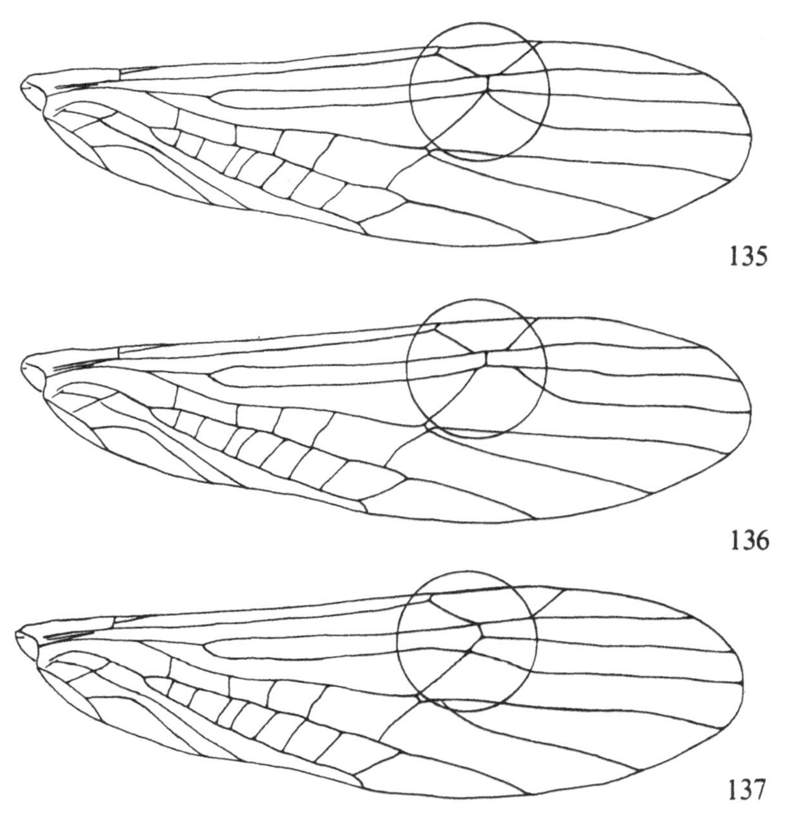

Figs 135-137. Different forms of wing venation seen in Nemouridae. – 135: the normal form with the typical X in the venation; 136 and 137 show variants.

The following synapomorphic characters are mentioned by Zwick (1973): the specialised internal structure of the male genitalia, the reduction in the number of abdominal ganglia, the flat saucer-like terminal segment of the labial palp, and the particular form and position of the front coxae.

The family consists of 14 genera, four of which occur in Fennoscandia and Denmark. The genera *Amphinemura* and *Protonemura* are separated from *Nemoura* and *Nemurella* by the presence of vestiges of nymphal gills on prosternum of the adult stonefly. Three sausage-like gills are present on each side of the nymphal prosternum in *Protonemura,* and two bunches, each of five to eight filamentous gills, in *Amphinemura* (Figs 138, 139).

Key to genera of Nemouridae

Adults

1 Vestiges of nymphal gills present on prosternum (Figs 138, 139) 2
- No vestiges of nymphal gills on prosternum 3

2(1) Three sausage-shaped prosternal gill-vestiges on each side (Fig. 139) . . . . *Protonemura* Kempny (p. 120)
- Two bunches of filamentous prosternal gill-vestiges on each side (Fig. 138) . . . . *Amphinemura* Ris (p. 91)

3(1) Male: cerci about as long as sternum 9 and modified into a pair of copulatory hooks (Figs 156-162); intermediate appendages reduced, subanal plates broad. Female: sternum 8 without lateral ridges, and anterior margin of sternum 9 elongate and triangular (Figs 179-185) . . . . *Nemoura* Latreille (p. 99)

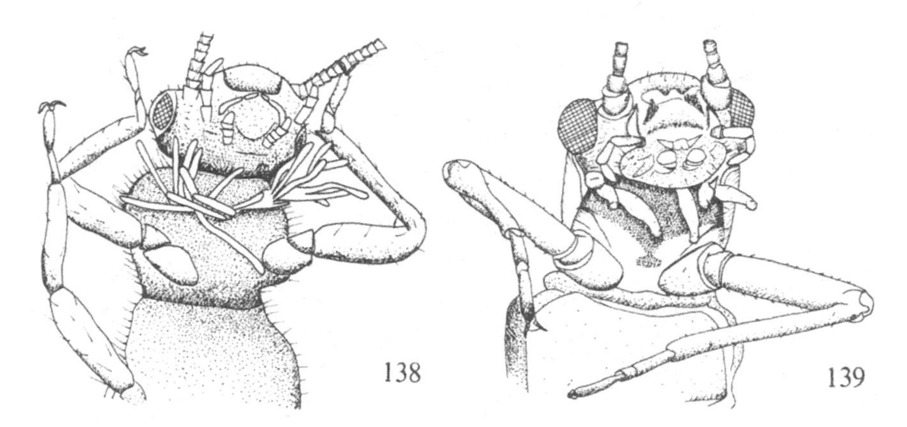

Figs 138, 139. Nymphal prosternum of 138: *Amphinemura* sp. with filamentous gills, and 139: *Protonemura* sp. with sausage-shaped gills.

- Male: cerci over two times as long as sternum 9 and without hooks; intermediate appendages membranous and digitiform, subanal plates narrow and leaf-shaped (Fig. 163). Female: sternum 8 with two lateral ridges, and anterior margin of sternum 9 straight (Fig. 186) .... *Nemurella* Kempny (p. 119)

Nymphs

1 Prosternal gills present 2
- Prosternal gills absent 3

2(1) Prosternum with three sausage-shaped gills on each side (Fig. 139) .... *Protonemura* Kempny (p. 120)
- Prosternum with two bunches of 5-8 filamentous gills on each side (Fig. 138) .... *Amphinemura* Ris (p. 91)

3(1) Segments 1 and 3 of hind tarsus of about equal length (Fig. 190). Hind femur with a well-defined transverse row of stout bristles (Fig. 190) .... *Nemurella* Kempny (p. 119)
- Segment 1 of hind tarsus about one third to one half as long as segment 3. Hind femur without transverse row of stout bristles .... *Nemoura* Latreille (p. 99)

Genus *Amphinemura* Ris, 1902

Nemoura (Amphinemura) Ris, 1902, Mitt. schweiz. ent. Ges., 10: 384.
Type species: *Nemoura cinerea* Olivier, 1811 (des. Claassen 1940).

Small species, dark brown to rufous brown in colour. Hind wings with a small anal area. Cerci short and simple. The male is easily recognised on the characteristic form of the epiproct (Figs 140-143), and the female on the shape of the subgenital plate (Figs 146-149). Body of nymphs covered with several thin bristles. Prosternum with a bunch of 5-8 filamentous gills on each side (Fig. 138).

Four species occur in Fennoscandia.

Key to species of *Amphinemura*

Adults

1 Epiproct present, males 2
- Epiproct absent, females 3

2(1) Epiproct cylindrical, rounded at apex. Inner lobe of subanal plate small (Fig. 140) .... 18. *borealis* (Morton)
- Epiproct cylindrical, pointed at apex. Inner lobe of subanal plate large (Fig. 143) .... 21. *sulcicollis* (Stephens)
- Epiproct knife-like, flat dorsally, sharp with several stout bristles ventrally. Inner and outer lobe of subanal plate well developed, rounded at apex (Figs 142, 144) .... 20. *standfussi* (Ris).

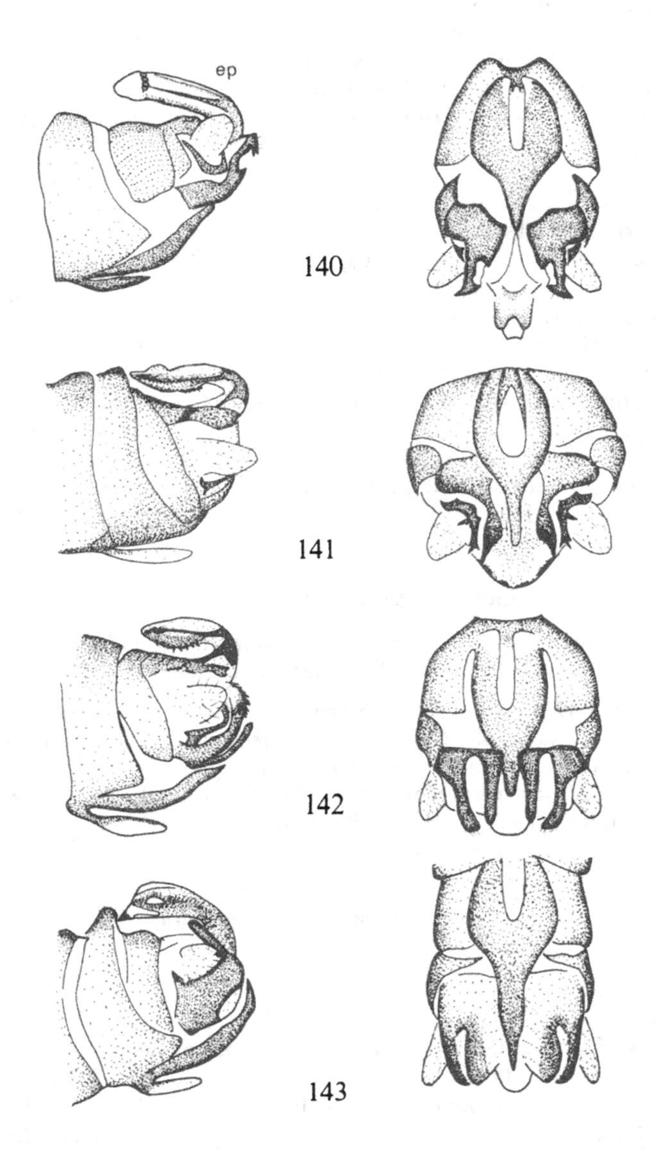

Figs 140-143. Male genitalia of *Amphinemura,* left row in lateral view, right row in ventral view. - 140: *A. borealis* (Mort.); 141: *A. palmeni* Kop.; 142: *A. standfussi* (Ris); 143: *A. sulcicollis* (Stph.); ep = epiproct.

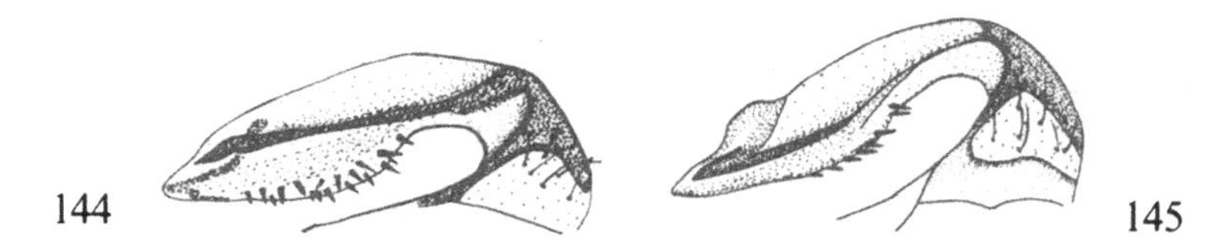

Figs 144, 145. Epiproct of 144: *Amphinemura standfussi* (Ris) and 145: *A. palmeni* Kop.

- Epiproct knife-like, with a hump dorsally, sharp with a few stout bristles ventrally. Inner lobe of subanal plate well developed and pointed at apex, outer lobe reduced (Figs 141, 145) .... 19. *palmeni* Koponen

3(1) Posterior part of sternum 7 large in middle, covering much of sternite 8. Subgenital plate unsclerotised, consisting of four lobes (Fig. 148) .... 20. *standfussi* (Ris)

- Posterior part of sternum 7 large in middle, covering much

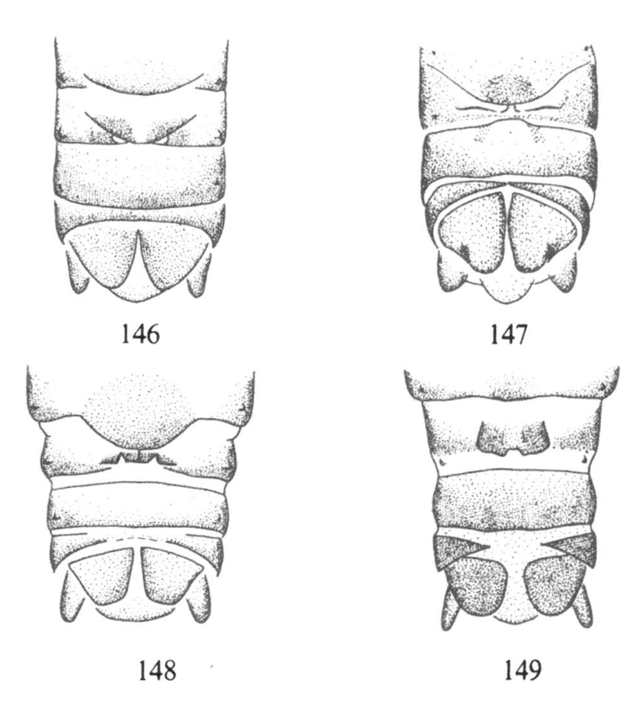

Figs 146-149. Female abdominal apex of *Amphinemura*, ventral view. - 146: *A. borealis* (Mort.); 147: *A. palmeni* Kop.; 148: *A. standfussi* (Ris); 149: *A. sulcicollis* (Stph.).

of sternite 8. Subgenital plate unsclerotised, consisting of two lobes (Fig. 147) .... 19. *palmeni* Koponen

- Posterior part of sternum 7 weakly developed. Subgenital plate sclerotised and triangular in shape (Fig. 146) ..... 18. *borealis* (Morton)
- Posterior part of sternum 7 weakly developed. Subgenital plate sclerotised and rectangular in shape (Fig. 149) . 21. *sulcicollis* (Stephens)

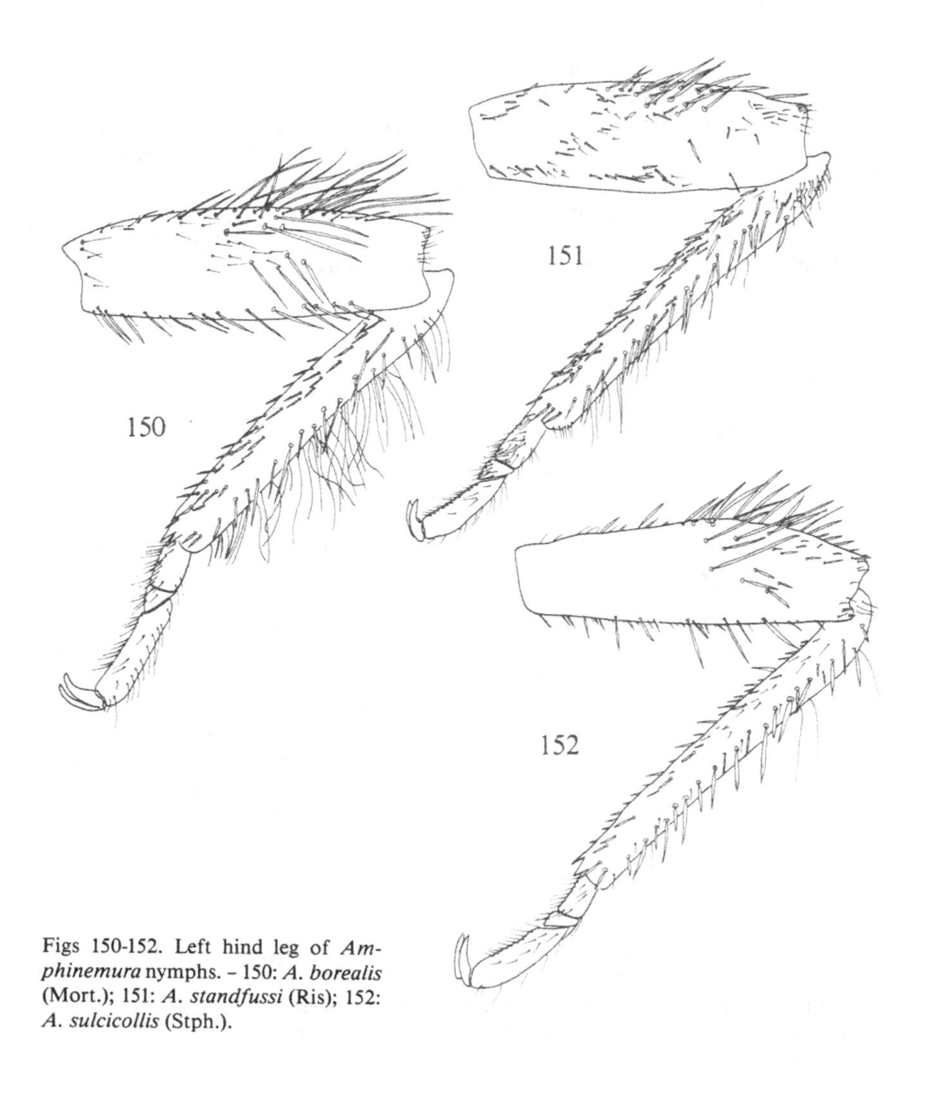

Figs 150-152. Left hind leg of *Amphinemura* nymphs. - 150: *A. borealis* (Mort.); 151: *A. standfussi* (Ris); 152: *A. sulcicollis* (Stph.).

Nymphs

1 Bristles on legs of two types: long and short (Fig. 150). Bristles on cercal segments 7-10 as long as or longer than segments (Fig. 153) .... 18. *borealis* (Morton)
- Bristles on legs mainly short and stout (Figs 151, 152). Bristles on cerci shorter than the segments (Figs 154, 155) 2

2(1) Cercal segments thickened towards apex (Fig. 155). Ventral surface of femora with both long and short bristles (Fig. 152).. 21. *sulcicollis* (Stephens)
- Cercal segments not thickened apically (Fig. 154). Ventral surface of femora with a few short bristles only (Fig. 151) ... 20. *standfussi* (Ris)

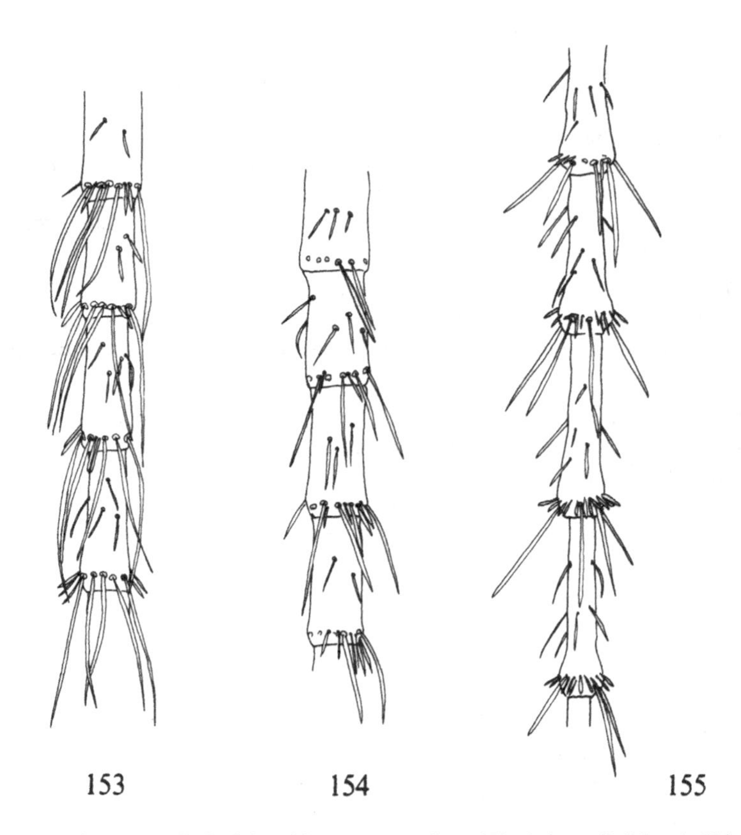

Figs 153-155. Cercal segments 7-10 of *Amphinemura* nymphs. - 153: *A. borealis* (Mort.); 154: *A. standfussi* (Ris); 155: *A. sulcicollis* (Stph.).

18. ***Amphinemura borealis*** (Morton, 1894)
Figs 29, 140, 146, 150, 153.

Nemoura borealis Morton, 1894, Trans. ent. Soc. Lond., 1894: 571.

Male. Body length 4.0-5.6 mm, wing length 6.1-7.2 mm. Small, slender and always long-winged. Body colour rufous brown to dark brown, wings light brown. Epiproct cylindrical, rounded at apex. Subanal plates broad and bilobed: inner lobe weakly developed, outer lobe large and dark pigmented with hooks at apex (Fig. 140).

Female. Body length 4.3-8.2 mm, wing length 7.5-9.5 mm. Colour similar to male. Sternite 7 elongate posteriorly. Subgenital plate on sternite 8 usually triangular and laterally well sclerotised and with lateral dark spots, subanal plates broad, pointed at apex (Fig. 146).

Nymphs. First instar 0.6-0.7 mm, fully grown nymphs 4.5-8.5 mm. Body rufous brown, and covered with short bristles. Both meso- and metanota anteriorly with a fringe of bristles. Legs covered with characteristically long, thin bristles, and tibiae with a fringe of long bristles (Fig. 150). Bristles of cerci as long, or longer, than the segments (Fig. 153).

Variability. The greatest variation is seen in the female, especially in regard to the shape of the subgenital plate which is generally triangular, but may also be rectangular and inserted posteriorly. The bristle arrangement on legs and cerci of nymphs may undergo some variation. However, *A. borealis* can usually be separated from the other species on the long bristles.

Distribution. *A. borealis* is absent from Denmark, but is distributed over most of Sweden, Norway and Finland. The species is supposed to be a north-eastern immigrant to Fennoscandia (Fig. 29). The species also occurs in central Europe, but not in Great Britain.

Biology. *A. borealis* occurs in both large and small streams, and has also been recorded from lakes in northern Fennoscandia. The species has a one-year life cycle, adults emerging in May-July. Occurs mainly in the boreal vegetation zone, rarely recorded from the subalpine zone. Often common in large rivers with a low water temperature during the summer.

19. ***Amphinemura palmeni*** Koponen, 1917
Figs 141, 145, 147.

Amphinemura palmeni Koponen, 1917, Acta Soc. Fauna Flora fenn., 44: 13.
Amphinemura norwegica Tobias, 1973, Senckenbergiana biol., 54: 339.

Male. Body length 4.3 mm, wing length 7.0 mm. Body colour dark brown to rufous brown. Epiproct has nearly the same shape as in *A. standfussi,* differences being the presence of a knob on the dorsal side, and of a few stout bristles ventrally (Figs 141, 145). Inner lobe of subanal plate strongly developed and pointed at apex, outer lobe reduced.

Female. Body length 5-7 mm, wing length 6-8 mm. Colour similar to male. Posterior part of sternite 7 greatly prolonged backwards in the middle. Subgenital plate unpigmented and bilobed. Subanal plate pointed at apex (Fig. 147).
Nymphs. Unknown.
Variability. There is some variability both in the male epiproct and in the female subgenital plate.

Distribution. Northern parts of Finland and Norway. The species was recently recorded by Thomas (1973) and Meinander (1980).

Biology. Adults recorded in July and August.

20. ***Amphinemura standfussi*** (Ris, 1902)
Figs 142, 144, 148, 151, 154.

Nemoura (Amphinemura) standfussi Ris, 1902, Mitt. schweiz. ent. Ges., 10: 395.
Nemoura nigra Zetterstedt, 1840, Ins. Lapp.: 1057 (preocc.).

Male. Body length 4.0-6.6 mm, wing length 3.3-6.5 mm. Body dark brown or rufous brown, wings light brown. Epiproct triangular, more or less knife-shaped, and with several short strong bristles. Inner lobe of subanal plate membraneous and rounded at apex, the outer part pointed, sclerotised, and with strong bristles at apex. Subgenital plate pointed at apex (Figs 142, 144).
Female. Body length 4.8-7.3 mm, wing length 2.8-7.3 mm. Colour similar to male. Posterior part of sternite 7 greatly prolonged backwards in the middle. Subgenital plate unpigmented, with four lobes. Subanal plates broad at apex (Fig. 148).
Nymphs. First instar 0.6-0.7 mm, fully grown nymph 4-8 mm. Body brown to rufous brown, head and legs chocolate brown. The ventral surface of all femora with short bristles only (Fig. 151). Cercal segments of uniform thickness, not widening towards apex, and bristles shorter than segments (Fig. 154).
Variability. Wide variations are found in this species, both in body length, wing length, short-wingedness and wing venation pattern. The variation in body and wing size are wider than in other *Amphinemura* species. In some localities entirely short-winged populations occur. Under special climatic conditions populations consisting of dwarf specimens occur. The shape of the genital appendages of both male and female also varies greatly (Lillehammer, 1974a, 1985b). The bristle arrangement on legs and cerci of nymphs show some variation, and single specimens, especially younger nymphs may be difficult to identify.

Distribution. *A. standfussi* occurs in Denmark, Norway, Sweden and Finland, but in absent from the south-western parts of both Sweden and Finland. The species may have reached Fennoscandia both from the south and from the north-east. The species occurs over most of Europe but only sporadically in the Mediterranean area.

Biology. *A. standfussi* occurs in both large rivers and small streams as well as in lakes. The species can be especially numerous in the outlet of high altitude lakes. *A.*

standfussi has a one-year life cycle, adults emerging in July-September. At high altitudes the eggs usually develop slowly during the winter, and hatching occurs in spring (Saltveit, 1977). The main nymphal growth period is during the summer. In lowland areas in the south, the eggs hatch in the autumn and nymphal growth takes place in early winter and in spring (Brinck, 1949).

21. *Amphinemura sulcicollis* (Stephens, 1836)
Figs 143, 149, 152, 155.

Nemoura sulcicollis Stephens, 1836, Illus. Brit. Ent., 6: 143.

Male. Small species. Body length 4.0-6.0 mm, wing length 5.7-6.9 mm. Colour dark brown to rufous brown, wings lighter brown with dark brown veins. Epiproct pointed at apex (Fig. 143). Subanal plate bilobed, the outer lobe dark pigmented, long and upwardly curved, the inner lobe membraneous and shorter. The subgenital plate with a much longer apex than in the other species.

Female. Body length 4.8-8.0 mm, wing length 6.1-8.8 mm. Body dark brown, wings light brown to smoky grey. The posterior part of sternite 7 only little prolonged backwards in the middle. Subgenital plate sclerotised and quadratic in shape, its colour brown. Subanal plate broad and rounded at apex (Fig. 149).

Nymphs. First instar 0.6-0.7 mm, fully grown nymph 4-9 mm. Body brown to rufous brown, head and legs often darker. Stout bristles on the legs. Cercal segments thickened towards the apex (Fig. 155). The ventral surface of the femora with both long and short bristles (Fig. 152).

Variability. The most pronounced variation occurs in the shape of the subgenital plate of the female, the male genitalia being fairly constant. Some variation is seen in the mid-wing venation pattern of both sexes (Lillehammer, 1974a). The bristle arrangement on legs and cerci of nymphs undergo some variation.

Distribution. *A. sulcicollis* occurs in all parts of Sweden and Norway; in Finland it is absent from the south-eastern part. In Denmark it occurs in Jutland but not on the islands. It may have entered Fennoscandia both from the south and the north-east. The species occurs over most of Europe including the Mediterranean area and Great Britain.

Biology. *A. sulcicollis* occurs in large rivers and small streams as well as in the exposed zone of lakes. The species has a one-year life cycle, adults emerging in May-July. The eggs develop rapidly at high water temperature and the main period of nymphal growth is in autumn and spring.

Genus *Nemoura* Latreille, 1796

Nemoura Latreille, 1796, Prec. caract. gén. Ins.: 101.
Type species: *Perla cinerea* Retzius, 1783 (des. Opinion 653 ICZN 1963).

Small to middle-sized species, larger than any species of *Amphinemura*. Gill vestiges absent in adults. Subgenital plate of male with a much smaller apex than in *Amphinemura*. Subanal plate usually unilobed. The female subgenital plate simple and less variable in form than in *Amphinemura*. The males of *Nemoura* species are mainly separated by the shape of the epiproct and the cerci, the females by the shape of the subgenital plate, and the structures beneath this. The nymphs are distinguishable by the arrangement of the bristles on pronotum, legs, abdominal terga and cercal segments. The first instar nymph has cerci composed of 3 segments, and antennae of 9 segments.

Key to species of *Nemoura*
(also including *Nemurella pictetii* Klap.)

Adults

1 Epiproct present, males 2
\- Epiproct absent, females 9
2(1) Cerci rounded at apex 3
\- Cerci with one or more hooks at apex 4
3(2) Epiproct divided in two or more parts (Fig. 171); subgenital plate with slender forked apex (Fig. 163) . . . 29. *Nemurella pictetii* Klapálek
\- Epiproct in one piece (Figs 167, 175); subgenital plate with an unforked apex (Fig. 159). . . . . 25. *Nemoura dubitans* Morton
4(2) Mid-part of epiproct elongated and pointed, extending beyond the anterior part 5
\- Mid-part of epiproct short, not extending beyond the anterior part 6
5(4) Cerci with two or three hooks. Epiproct (Figs 169, 177). Subanal plate curved posteriorly (Fig. 161) . 27. *Nemoura sahlbergi* Morton
\- Cerci with one, rarely two hooks. Epiproct (Figs 166, 174). Subanal plate pointed with an insertion posteriorly (Fig. 158). . . . . 24. *Nemoura cinerea* (Retzius)
6(4) Mid-part of epiproct blunt anteriorly 7
\- Mid-part of epiproct pointed anteriorly 8
7(6) Mid-part of epiproct broad, blunt anteriorly (Figs 164, 172). Cerci with two or more hooks (Fig. 156) . . . . 22. *Nemoura arctica* Esben-Petersen
\- Mid-part of epiproct broad and darkly pigmented, toothed at apex (Figs 165, 173). Cerci with one hook (Fig. 157). . . . . 23. *Nemoura avicularis* Morton

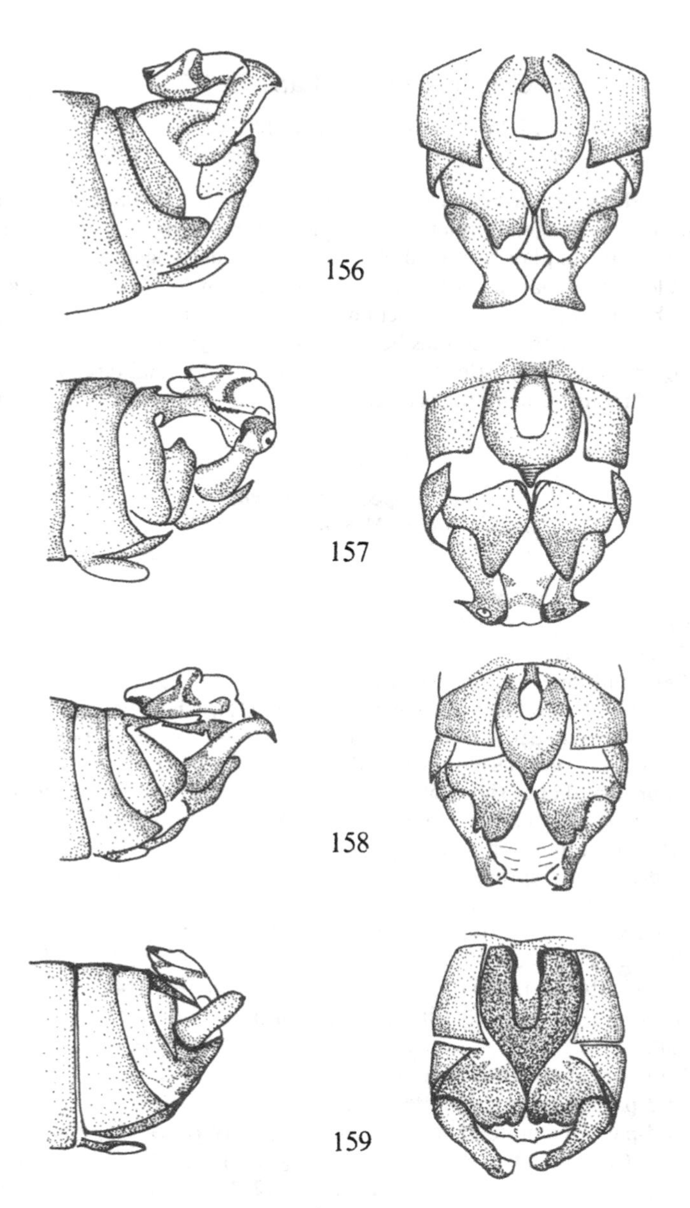

Figs 156-159. Male abdominal apex of *Nemoura*, left row in lateral view, right row in ventral view. - 156: *N. arctica* Esben-P.; 157: *N. avicularis* Mort.; 158: *N. cinerea* (Retz.); 159: *N. dubitans* Mort.

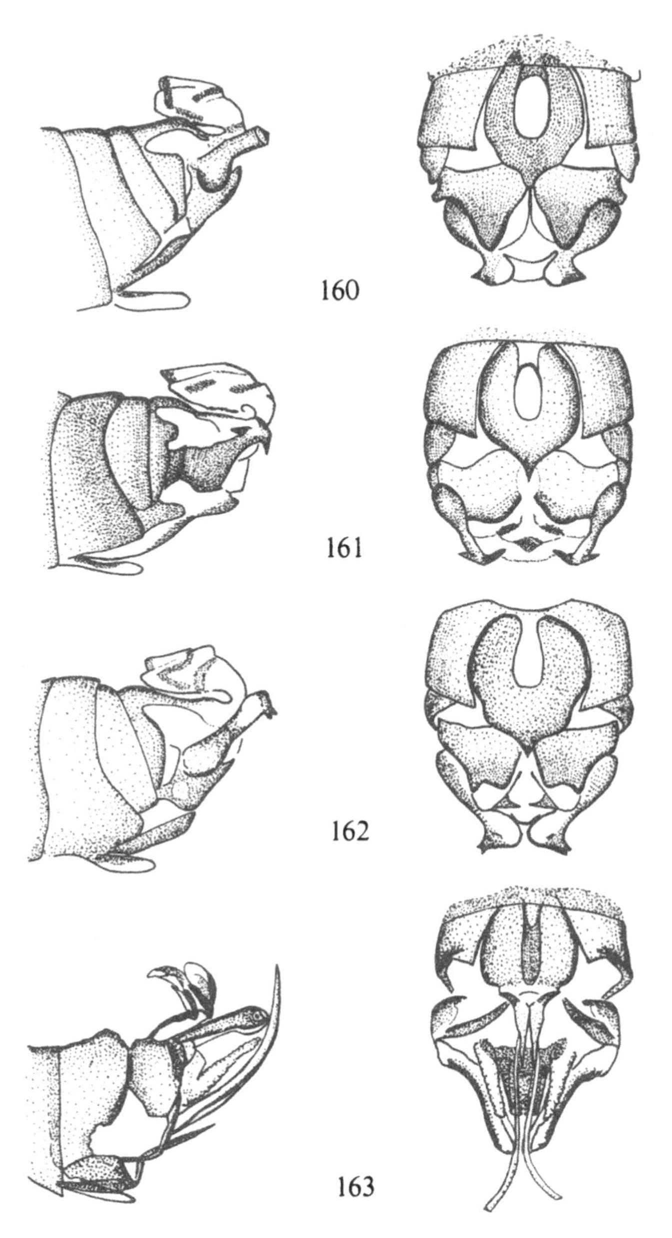

Figs 160-163. Male abdominal apex of *Nemoura* and *Nemurella*, left row in lateral view, right row in ventral view. - 160: *N. flexuosa* Aub.; 161: *N. sahlbergi* Mort.; 162: *N. viki* Lilleh.; 163: *Nemurella pictetii* Klap.

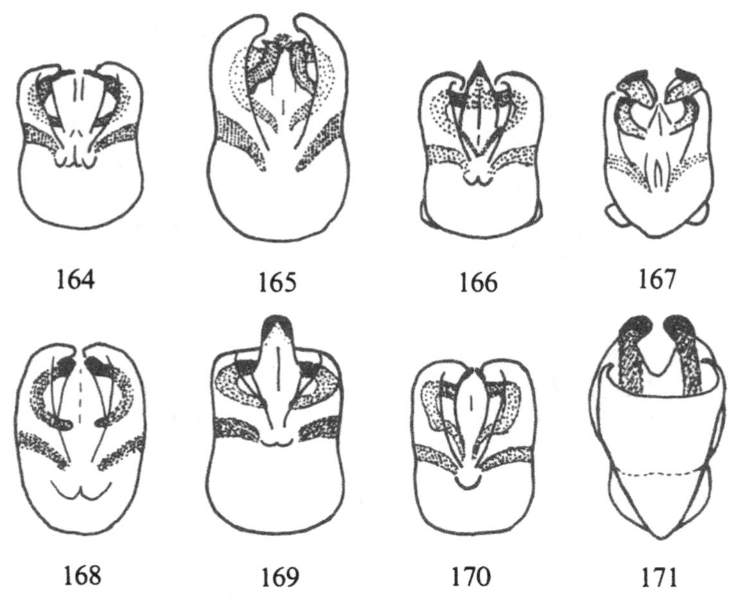

Figs 164-171. Male epiproct of *Nemoura* and *Nemurella*, dorsal view. - 164: *N. arctica* Esben-P.; 165: *N. avicularis* Mort.; 166: *N. cinerea* (Retz.); 167: *N. dubitans* Mort.; 168: *N. flexuosa* Aub.; 169: *N. sahlbergi* Mort.; 170: *N. viki* Lilleh.; 171: *Nemurella pictetii* Klap.

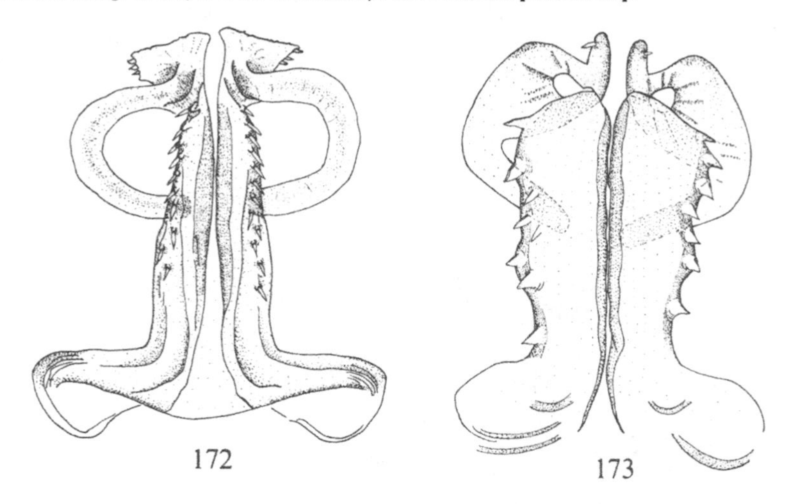

Figs 172-178. Inner structure of male epiproct of *Nemoura*, seen from below. - 172: *N. arctica* Esben-P.; 173: *N. avicularis* Mort.; 174: *N. cinerea* (Retz.); 175: *N. dubitans* Mort.; 176: *N. flexuosa* Aub.; 177: *N. sahlbergi* Mort.; 178: *N. viki* Lilleh.

8(7) Cerci with one hook. Inner sclerotised part of epiproct rounded and dark-pigmented (Figs 168, 176) . 26. *Nemoura flexuosa* Aubert

\- Cerci with two or more hooks (Fig. 162). Inner sclerotised part of epiproct rectangular (Figs 170, 178) . 28. *Nemoura viki* Lillehammer

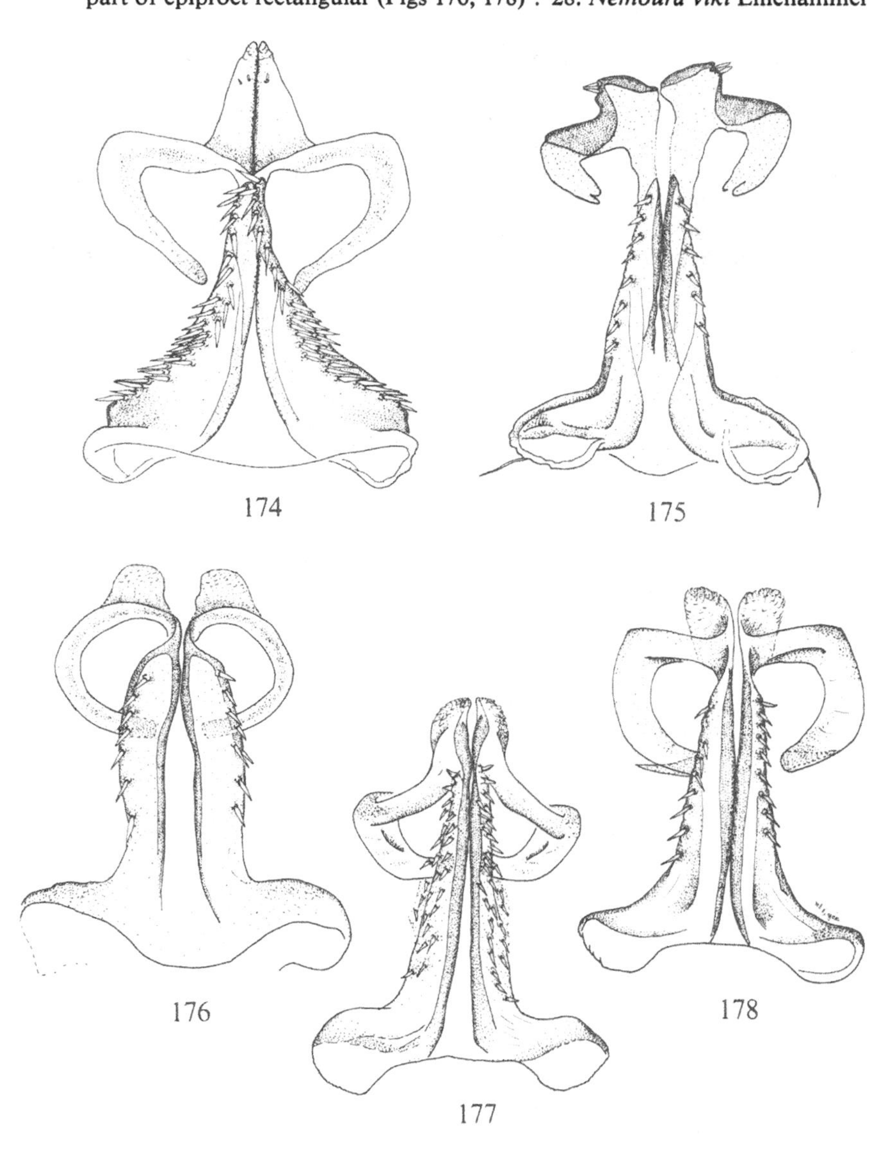

9(1) Structures present beneath subgenital plate 10
- No structures present beneath subgenital plate 13
10(9) Sclerotised and pigmented plates present beneath sub-genital plate 11
- Unpigmented structures present beneath subgenital plate 12
11(10) Three sclerotised plates present beneath subgenital plate (Fig. 182) .... 25. *Nemoura dubitans* Morton
- Two small sclerotised plates present in the middle beneath subgenital plate (Fig. 183) .... 26. *Nemoura flexuosa* Aubert
12(10) A three-segmented unpigmented structure present below subgenital plate (Fig. 180) .... 23. *Nemoura avicularis* Morton
- A two-segmented unpigmented structure present beneath subgenital plate (Fig. 181) .... 24. *Nemoura cinerea* (Retzius)

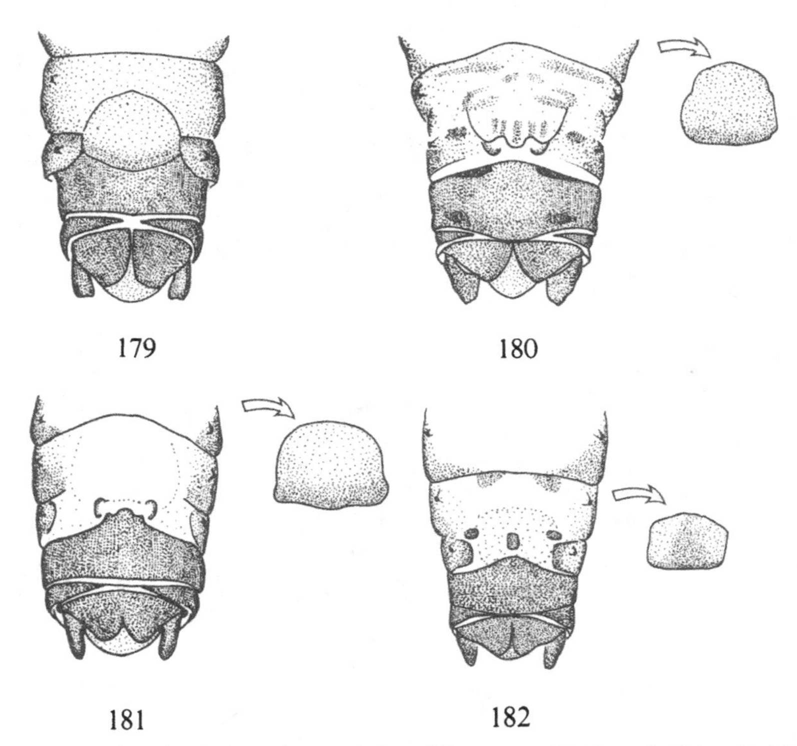

Figs 179-182. Female abdominal apex in ventral view of *Nemoura*. - 179: *N. arctica* Esben-P.; 180: *N. avicularis* Mort.; 181: *N. cinerea* (Retz.); 182: *N. dubitans* Mort. - In 180-182 is the subgenital plate arranged to the right in order to show the structures beneath it.

13(9) Subgenital plate more or less triangular (Fig. 186). Two large, laterally sclerotised protuberances on sternite 8 .... 29. *Nemurella pictetii* Klapálek

\- Subgenital plate rounded posteriorly 14

14(13) Mid-part of subgenital plate with posterior projection (Fig. 184) .... 27. *Nemoura sahlbergi* Morton

\- Mid-part of subgenital plate truncate or rounded 15

15(14) Sternum 7 heavily pigmented laterally (Fig. 179) .... 22. *Nemoura arctica* Esben-Petersen

\- Sternum 7 unpigmented (Fig. 185) .... 28. *Nemoura viki* Lillehammer

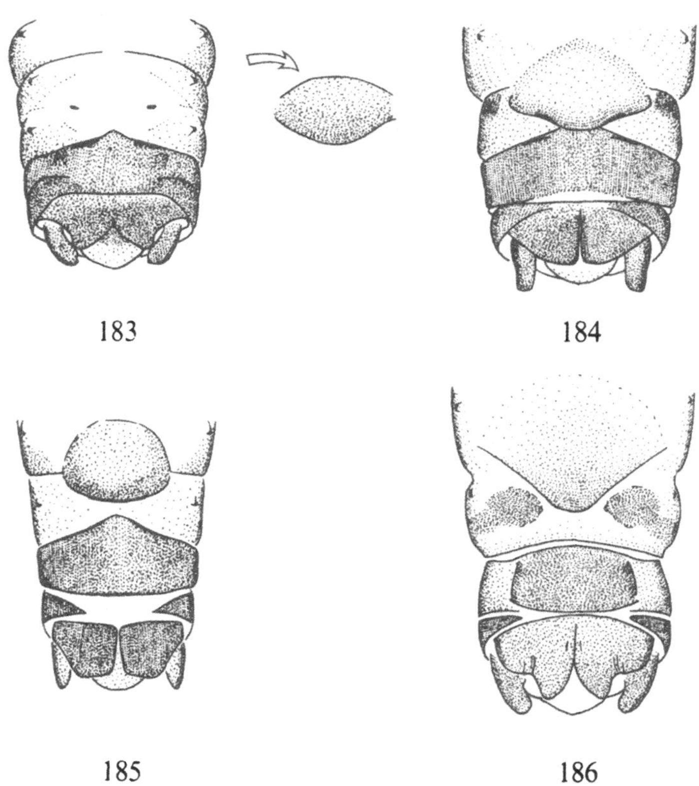

Figs 183-186. Female abdominal apex in ventral view of *Nemoura* and *Nemurella*. - 183: *N. flexuosa* Aub.; 184: *N. sahlbergi* Mort.; 185: *N. viki* Lilleh.; 186: *Nemurella pictetii* Klap. - In 183 is the subgenital plate arranged to the right in order to show the structures beneath it.

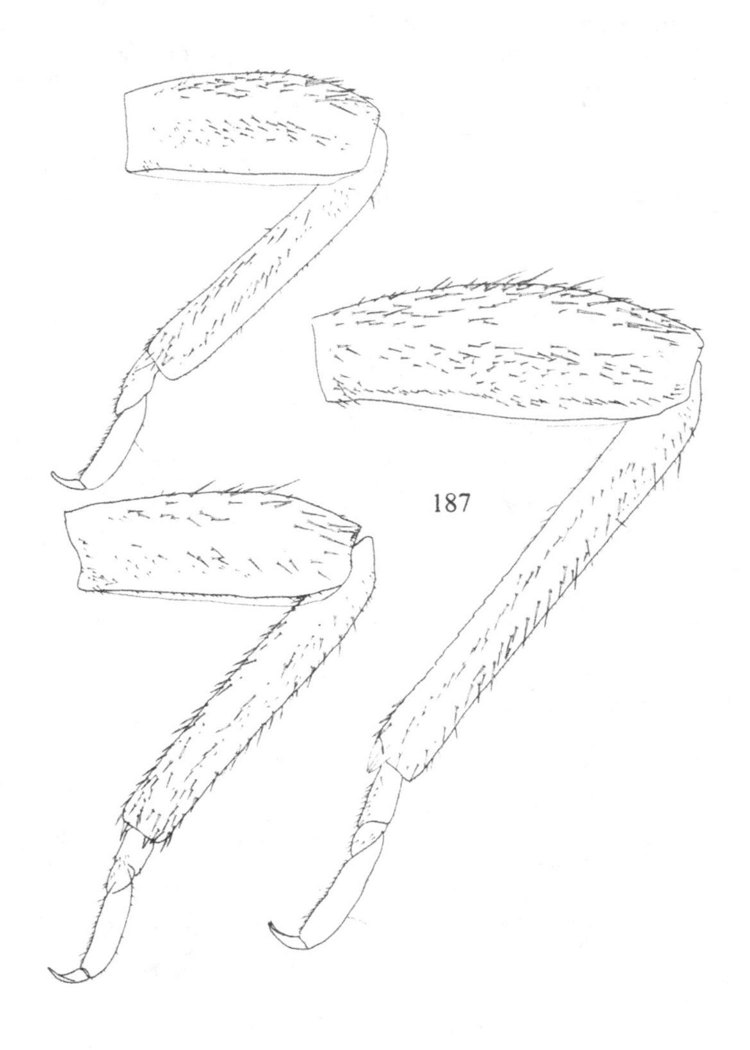

Figs 187-189. Fore, mid and hind legs of nymphs of *Nemoura*. - 187: *N. arctica* Esben-P.; 188: *N. avicularis* Mort.; 189: *N. flexuosa* Aub.
Fig. 190. Hind leg of *Nemurella pictetii* Klap.

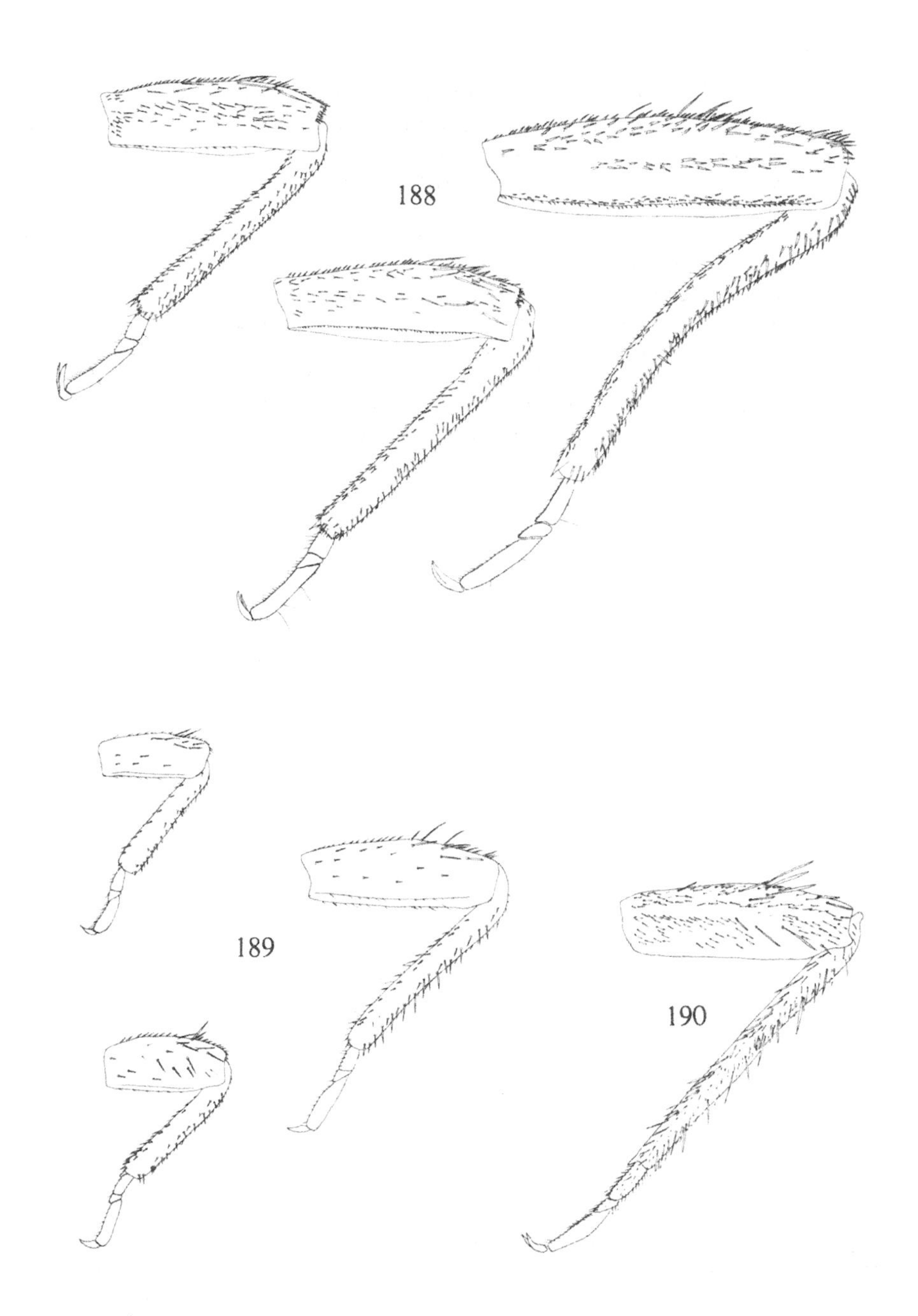
188
189
190

Nymphs

1	Femora of one or more legs with discrete long bristles distally; tibiae mainly with short stout bristles (Figs 188, 189)	2
-	Femora of all three legs with bristles of different length scattered over surface; tibiae with bristles of different shape (Fig. 187)	5
2(1)	Femora 1 and 2 with both long and short bristles	3
-	All three femora with both long and short bristles	4
3(2)	Bristles of cercal segments 15-17 shorter than half segmental length, a few surface bristles present (Fig. 193). Abdominal tergites 4-7 with a fringe of bristles that are shorter than half tergal length (Fig. 198)	24. *Nemoura cinerea* (Retzius)
-	Bristles of cercal segments 15-17 of about same length as the segments, long and stout surface bristles present (Fig.	

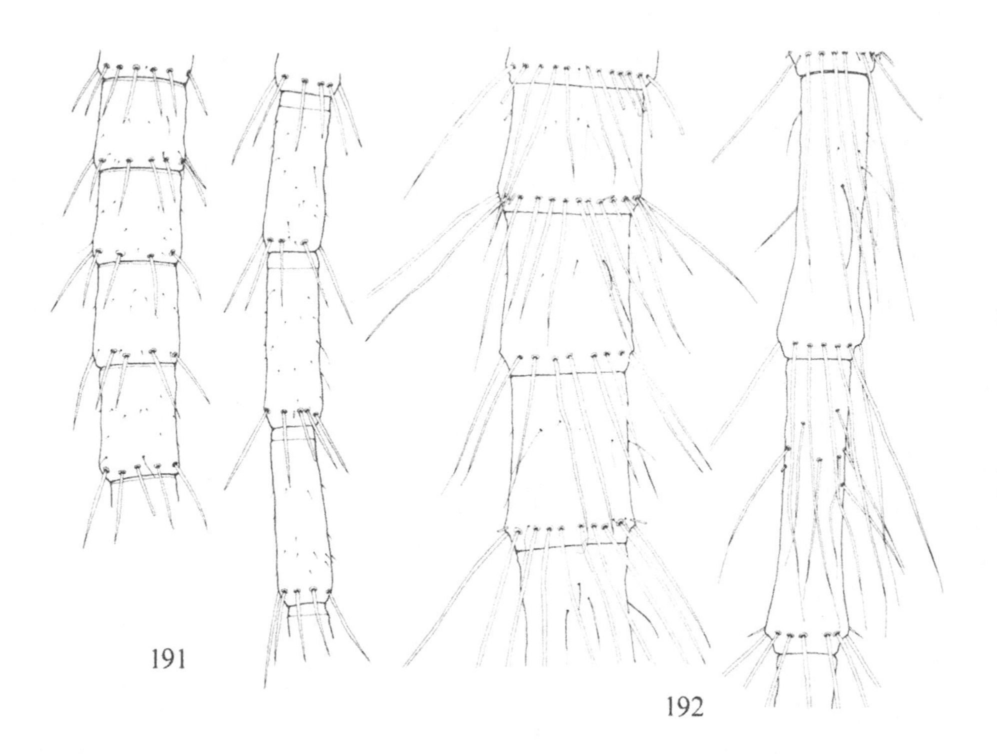

Figs 191, 192. Cercal segments 7-10 and 15-17 of *Nemoura* nymphs. - 191: *N. arctica* Esben-P.; 192: *N. avicularis* Mort.

192). Abdominal tergites 4-7 with both long and short bristles along posterior margin (Fig. 199) .. 23. *Nemoura avicularis* Morton

4(2) First tarsal segment curved (Fig. 203). Cercal segments with stout bristles longer than half segmental length; surface bristles present on segments 7-10, as well as on 15-17 (Fig. 194). Abdominal tergites 4-7 with very short bristles along posterior margin (Fig. 200) .... 25. *Nemoura dubitans* Morton

\- First tarsal segment straight (Fig. 202). Cercal segments with stout bristles shorter than half segmental length; no surface bristles present on segments 7-10 (Fig. 195). Abdominal tergites 4-7 with a fringe of bristles shorter than half tergal length (Fig. 201) .... 26. *Nemoura flexuosa* Aubert

5(1) Galea with hairs (Fig. 205) 6

\- Galea without hairs 7

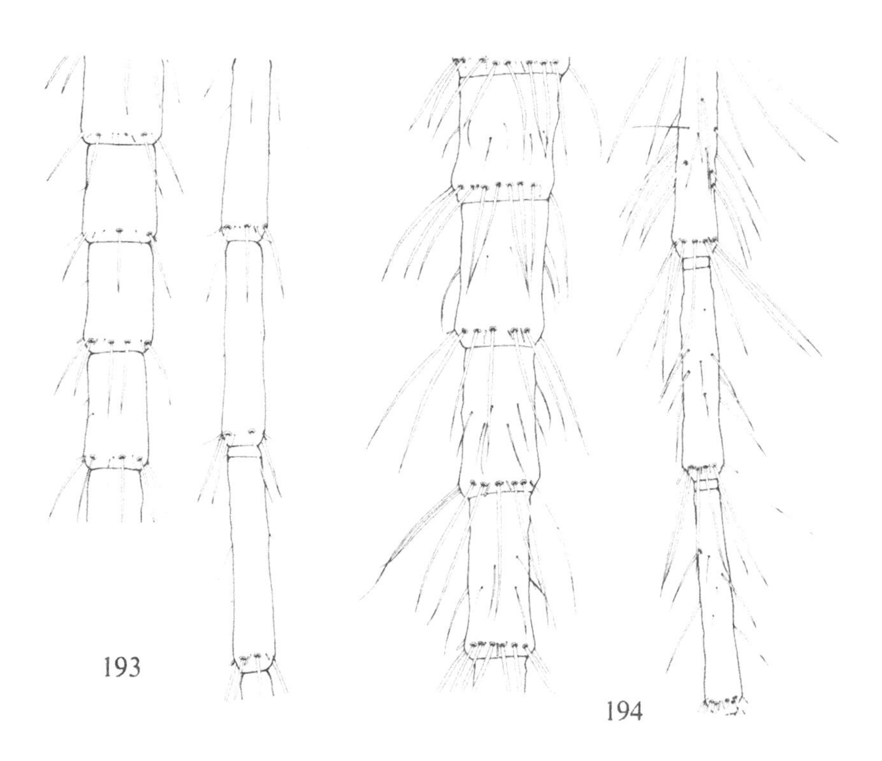

Figs 193, 194. Cercal segments 7-10 and 15-17 of *Nemoura* nymphs. - 193: *N. cinerea* (Retz.); 194: *N. dubitans* Mort.

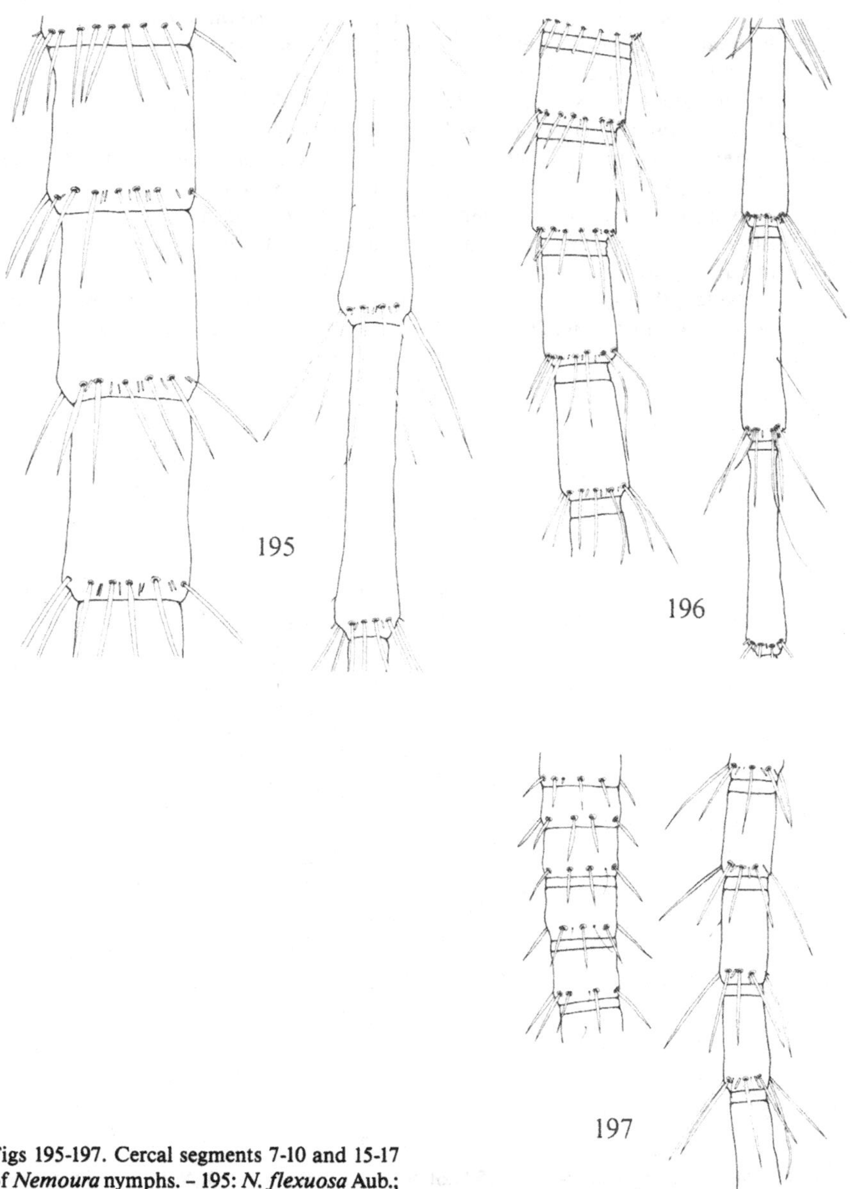

Figs 195-197. Cercal segments 7-10 and 15-17 of *Nemoura* nymphs. – 195: *N. flexuosa* Aub.; 196: *N. sahlbergi* Mort.; 197: *N. viki* Lilleh.

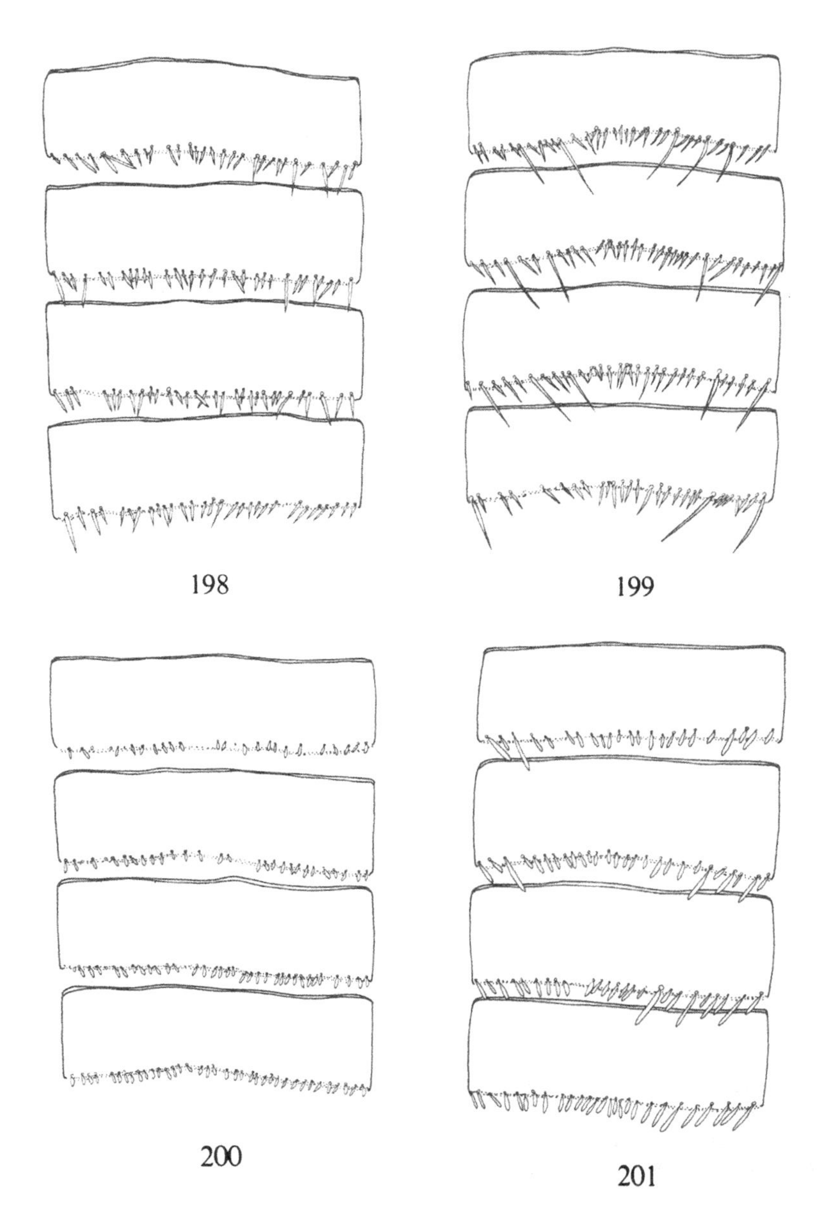

Figs 198-201. Abdominal terga 4-7 of nymphs of *Nemoura*. – 198: *N. cinerea* (Retz.); 199: *N. avicularis* Mort.; 200: *N. dubitans* Mort.; 201: *N. flexuosa* Aub.

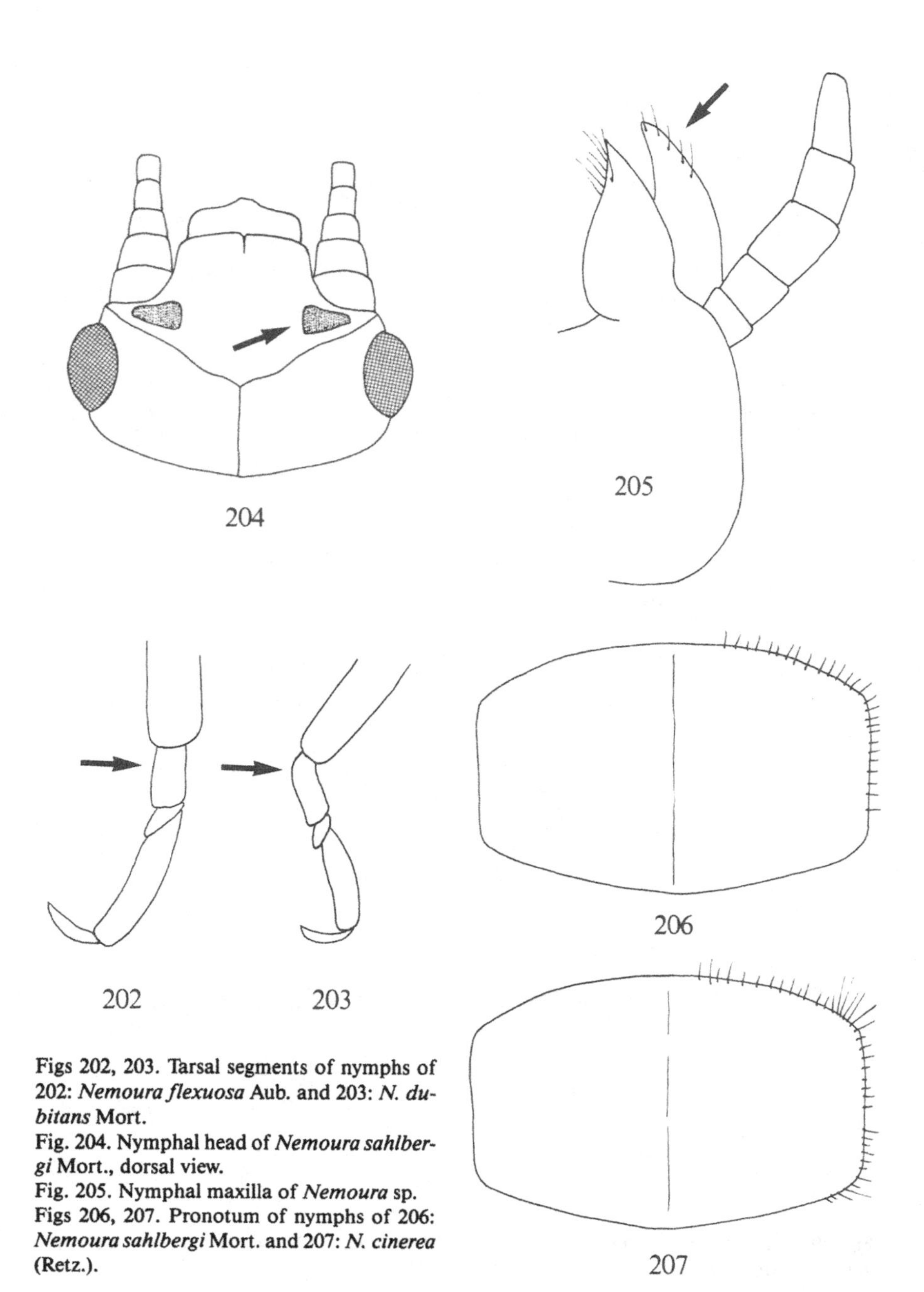

Figs 202, 203. Tarsal segments of nymphs of 202: *Nemoura flexuosa* Aub. and 203: *N. dubitans* Mort.
Fig. 204. Nymphal head of *Nemoura sahlbergi* Mort., dorsal view.
Fig. 205. Nymphal maxilla of *Nemoura* sp.
Figs 206, 207. Pronotum of nymphs of 206: *Nemoura sahlbergi* Mort. and 207: *N. cinerea* (Retz.).

6(5) Pronotum with a marginal fringe of short bristles (as in Fig. 206). Bristles on cercal segments 15-17 longer than half segmental length; no surface bristles present on segments 7-10 (Fig. 197) .... 28. *Nemoura viki* Lillehammer

– Pronotum with a marginal fringe of both long and short bristles (Fig. 207). Bristles on cercal segments 15-17 shorter than half segmental length; surface bristles present on segments 7-10 (Fig. 193) .... 24. *Nemoura cinerea* (Retzius)

7(5) Pronotum with a marginal fringe of short bristles only (Fig. 206). Some bristles on cercal segments 7-10 nearly as long as, or longer than, half segmental length; no surface bristles present on segments 7-10 (Fig. 196) . 27. *Nemoura sahlbergi* Morton

– Pronotum with a marginal fringe of both long and short bristles (as in Fig. 207). Bristles on cercal segments 7-10 about as long as the segment; surface bristles present on segments 7-10 and 15-17 (Fig. 191) .... 22. *Nemoura arctica* Esben-Petersen

22. ***Nemoura arctica*** Esben-Petersen, 1910
Figs 31, 156, 164, 172, 179, 187, 191.

Nemoura arctica Esben-Petersen, 1910a, Tromsø Mus. Aarsh., 31/32: 85.

Male. Body length 5.4-7.0 mm. Wing length 4.9-7.3 mm. Body dark brown, wings light brown. Middle part of epiproct blunt and rounded at apex. Subanal plates broad, varying in form; in some populations bilobed, in others with a single lobe (Lillehammer, 1974a). The cerci possess two or more hooks; also cerci show large variations (Figs 156, 164, 172).

Female. Body length 6.5-9.6 mm. Wing length 6.0-8.6 mm. Body colour similar to male. The subgenital plate dark coloured and rounded posteriorly. Abdominal segment 7 heavily pigmented (Fig. 179).

Nymphs. First instar 0.58-0.67 mm. Full-grown nymphs 6.0-9.0 mm. Reddish brown in colour, darker on the dorsal side. Head with kidney-shaped dark spots behind antennae, ocelli invisible. Galea without hairs. First two segments of antennae may sometimes be darker than the remainder. Pronotum with a marginal fringe of both long and short bristles (as in Fig. 207). All femora with mainly short bristles (Fig. 187). Abdominal tergites have mainly short bristles along posterior margin. Cercal segments with a fringe of stout bristles posteriorly and a dense cover of surface bristles (Fig. 191).

Variability. A high degree of variation occurs in nearly all the characters of the genital appendages of both sexes, as well as in the wing venation pattern, wing length and body length. Short-winged populations occur in some localities. Within its overall distributional area in Fennoscandia this species seems capable of forming local populations that are morphologically distinct (Lillehammer, 1974a). Outside Europe some subspecies have been described.

Distribution. This circumpolar species has only been found in the northern parts of Finland, Norway and Sweden, and is a north-eastern immigrant to Fennoscandia (Fig. 31).

Biology. *N. arctica* occurs in both small and large streams and in lakes. Under favourable conditions the species seems to have a one-year life cycle but under unfavorable temperature conditions this may change to a two-year life cycle. The incubation period of the eggs is short at high and long at low water temperatures (Lillehammer, 1986b). Emergence in June-July. *N. arctica* may often be the sole stonefly species found in small streams at high altitudes.

23. ***Nemoura avicularis*** Morton, 1894
Figs 157, 165, 173, 180, 188, 192, 199.

Nemoura avicularis Morton, 1894, Trans. ent. Soc. Lond., 1894: 562.

Male. Body length 5.8-8.0 mm, wing length 6.8-8.8 mm. Middle part of epiproct small, cerci possessing only one hook (Figs 157, 165, 173). Subanal plate with one lobe, subgenital plate broad with a small apex. Body dark brown, wings smoky in colour.

Female. Body length 6.8-10.0 mm, wing length 8.0-12.0 mm. Abdominal segment 8 has a three-lobed unpigmented structure beneath the subgenital plate (Fig. 180). Body dark brown, wings lighter, with dark brown veins.

Nymphs. First instar about 0.70 mm. Fully grown nymph 6.0-10.0 mm. Body usually greyish brown. Head with two elongate, darker spots, ocelli invisible. Galea with hairs (Fig. 205). First two segments of antennae always darker than the remainder, all segments have short bristles. Pronotum with a dense marginal fringe of short bristles (as in Fig. 206). Femora of first two legs usually with both long and short bristles (Fig. 188). Abdominal tergites with both long and short bristles (Fig. 199). Cerci long and slender, with a dense fringe of long apical bristles on each segment and also with long intermediate bristles on the segments (Fig. 192).

Variability. Some variations occur in the internal structure of the epiproct and the subgenital plate of the male. Variation in the wing venation is found in both sexes, but individuals of both sexes are always long-winged. There is some variation in the bristle arrangement on the legs, viz., sometimes the femora of all three legs have both long and short bristles, while at other times only the fore and mid femora have bristles of different length. The variation in the taxonomical characters of nymphs seems to be greater in the Fennoscandian specimens than that reported from Great Britain (Hynes 1963).

Distribution. Occurs over most of Fennoscandia and Denmark, but is absent from some of the Danish islands. The species is not common in western Norway, and may have entered Fennoscandia both from the south and from the north-east. *N. avicularis* occurs over most of Europe, except the Mediterranean, eastward to Siberia.

Biology. *N. avicularis* occurs in small streams and large rivers as well as in the exposed zones of lakes. It may also occur at habitats with fine substratum and much dead organic material. The species has a one-year life cycle, and adult emergence takes place in April-July. Nymphal growth mainly occurs in late summer and autumn (Brittain, 1974; Lillehammer, 1978).

24. ***Nemoura cinerea*** (Retzius, 1783)
Figs 158, 166, 174, 181, 193, 198, 207.

Perla cinerea Retzius, 1783, Car. de Geer gen. et spec.: 60.
Nemoura nebulosa Stephens, 1836, Illus. Brit. Ent., 6: 140.
Nemoura cruciata Stephens, 1836, Illus. Brit. Ent., 6: 141.
Nemoura fuliginosa Stephens, 1836, Illus. Brit. Ent., 6: 141.
Nemoura pusilla Stephens, 1836, Illus. Brit. Ent., 6: 142.
Nemoura umbrosa Pictet, 1865, Névr. d'Espagne: 20.
Nemoura variegata auctt., nec Olivier.

Male. Body length 4.8-9.0 mm, wing length 4.5-8.6 mm. Body dark red-brown, wings light brown. The only species of *Nemoura* with a punctured pronotum. Middle part of epiproct large (Figs 158, 166, 174), exceeding the anterior part in length. Subanal plates with one large and one small lobe. Subgenital plate with a small apex. Cerci with one, rarely two hooks.

Female. Body length 5.3-11.0 mm, wing length 5.4-9.9 mm. Colour similar to male, wings light brown. Pronotum punctured. Two unpigmented structures beneath the subgenital plate (Fig. 181).

Nymphs. Body light brown to dark reddish brown. First instar 0.6-0.7 mm. Fully-grown nymphs 4.0-9.0 mm. Three ocelli may or may not be visible. Dark patches on head triangular. First antennal segments may be darker than the remainder, though there is a wide variation. Galea with hairs (Fig. 205). Pronotum with a sparse marginal fringe of long and short bristles (Fig. 207). Legs with a variable covering of bristles. Abdominal tergites with both long and short bristles (Fig. 198). Cercal segments with a fringe of short bristles apically, and a few short and thin bristles present on the surface of the segments (Fig. 193). Abdominal terga 4-6 with bristles shorter than half length of segments.

Variability. Some variation occurs in the form of the cercal hooks, in the shape of the subanal and subgenital plates of the male, and in the form of the female subgenital plate (Lillehammer, 1974a). Marked variation also occurs in the taxonomical characters of the nymphs. Ocelli may be present in specimens from some populations, but not in others. Some specimens may have bristles of different length on all femora, although most often the hind femur has short hairs of even length.

Distribution. *N. cinerea* occurs in all parts of Fennoscandia and in Denmark, where it is also common on the islands. The species may have dispersed into Fennoscandia both from the south and the north-east. This species occurs all over Europe, including Great Britain and the Mediterranean area, eastward to Central Asia.

Biology. *N. cinerea* occurs in all types of freshwater bodies: small streams, large rivers, small and large lakes, and sometimes also in ponds. The species has a one-year life cycle under favourable conditions, but a two-year life cycle under unfavourable conditions (Brittain, 1974; Lillehammer, 1975b). Emergence from May to September. Nymphal growth may be unsyncronized in some habitats.

25. ***Nemoura dubitans*** Morton, 1894
Figs 159, 167, 175, 182, 194, 200, 203.

Nemoura dubitans Morton, 1894, Trans. ent. Soc. Lond., 1894: 565.

Male. Body length 6.5-7.5 mm, wing length 7.2-8.3 mm. Body dark brown, legs red-brown, wings light grey with darker veins; only long-winged specimens recorded. Epiproct small (Figs 159, 167, 175). Subgenital plate with a pointed apex, subanal plates with a blunt apex. Cerci rounded posteriorly, without hooks.

Female. Body length 7.5-9.0 mm, wing length 8.1-9.4 mm; only long-winged specimens recorded. Colour similar to male. Three sclerotised plates present beneath the subgenital plate which is small (Fig. 182).

Nymphs. Fully grown nymph 6-9 mm. Body grey-brown. Head with kidney-shaped dark patches, ocelli (if visible) unpigmented. Galea processes hairs (as in Fig. 205). Pronotum with a dense marginal fringe of short and stout bristles (as in Fig. 206). All femora with bristles of different length. Tibiae have short stout bristles along the margin. First tarsal segment curved (Fig. 203). Abdominal terga with short bristles only (Fig. 200). Cercal segments with a dense fringe of long bristles apically, and long and stout bristles present also on the surface of the segments (Fig. 194).

Variability. Too few specimens have been examined for any conclusion to be drawn.

Distribution. In Denmark *N. dubitans* occurs in the eastern part of Jutland and on Zealand. In Sweden it occurs only in the south, and in Finland only in the south-western part. *N. dubitans* has not been recorded from Norway. The species occurs in central and northern Europe, including parts of Great Britain.

Biology. Little is known about the biology of this species in Fennoscandia, but Hynes (1977) states that in Great Britain nymphs inhabit small, shallow, overgrown, spring-fed streams. In Sweden emergence occurs in April-May.

26. ***Nemoura flexuosa*** Aubert, 1949
Figs 160, 168, 176, 183, 189, 195, 201, 202.

Nemoura flexuosa Aubert, 1949, Mitt. schweiz. ent. Ges., 22: 218.
Nemoura erratica s. auctt., nec Claassen, 1936.

Male. Body length 4.8-8.0 mm, wing length 6.0-8.0 mm. Body dark brown, wings light brown with darker veins. Middle part of epiproct not overreaching the anterior part

(Figs 160, 168, 176). Subgenital plate broad, with a small apex, subanal plates with only one lobe. Cerci with a single hook.

Female. Body length 5.3-10.0 mm, wings 6.8-10.3 mm. Colour similar to male. Subgenital plate small; two small sclerotised plates are placed laterally beneath it (Fig. 183).

Nymphs. Fully-grown nymphs 6-9 mm. Body dark reddish brown. Galea possesses hairs. Head with dark-coloured crescent-shaped spots. Pronotum has a dense marginal fringe of short bristles. Femora of all legs have bristles of different length (Fig. 189). Tibiae have short, stout bristles along the margin. First tarsal segment straight (Fig. 202). Abdominal terga with a fringe of short blunt bristles (Fig. 201). Cercal segments with a fringe of long stout bristles apically, bristles being shorter than length of the individual segments. Bristles absent from the surface of the segments (Fig. 195).

Variability. Small variations occur in both the shape of the male genital appendages and in the wing venation, as well as in the bristle arrangement on legs and cerci of nymphs.

Distribution. *N. flexuosa* occurs in Sweden, Finland, Denmark and Norway. In the western parts of Denmark and Norway it is rare or may be entirely absent. In Sweden and Finland it is rare in the north, whereas it is common in northern Norway. The species is supposed to be a southern immigrant to Fennoscandia. *N. flexuosa* occurs in northern, eastern and central Europe and in Italy, but not in Great Britain and Spain.

Biology. *N. flexuosa* occurs mainly in small streams. Little is known about the life cycle, but it is supposed to last one year, adult emergence taking place in April-June in the south and in July in the north.

27. ***Nemoura sahlbergi*** Morton, 1896

Figs 24, 161, 169, 177, 184, 196, 204, 206.

Nemoura sahlbergi Morton, 1896, Trans. ent. Soc. Lond., 1896: 56.

Male. Body length 3.8-6.2 mm, wing length 5.7-7.2 mm. Body dark rufous brown, wings light brown. Middle part of epiproct greatly overreaching the anterior part (Figs 161, 169, 177). Cerci with two or three hooks. Subanal plates curved posteriorly (Fig. 161). Subgenital plate broad, with a small apex.

Female. Body length 5.0-8.0 mm, wing length 6.2-7.6 mm. Colour similar to male. The heavily pigmented subgenital plate pointed posteriorly (Fig. 184).

Nymphs. First instar nymph about 0.6 mm. Fully grown nymph 6.0-7.0 mm. Body ochre-brown to dark brown. Head with two triangular dark-pigmented markings (Fig. 204). First three antennal segments often darker than rest. Galea without hairs. Pronotum with a sparse fringe of short bristles (Fig. 206). All femora with short bristles only on the posterior margins. Abdominal terga 4-6 with a fringe of bristles; these about half as long as tergal length. Tibiae with long and thin bristles along the margin. Cercal segments apically with bristles as long as length of the individual segment; only a few bristles present on the surface of the segment (Fig. 196). The nymph was described by Lillehammer (1972b).

Variability. Wide variations occur in the form of the subanal plates and of the cercal hooks of the male, and also in the form of the female subgenital plate. Some variation may also occur in the venation pattern and in the wing length (Lillehammer, 1974a).

Distribution. *N. sahlbergi* occurs only in the northernmost part of Fennoscandia; not found in Denmark. The species is a north-eastern immigrant to our area and probably has a circumpolar distribution (Fig. 24).

Biology. *N. sahlbergi* occurs mainly in small streams in the subalpine vegetation zone and has a one-year life cycle, adults emerging in June and July.

28. ***Nemoura viki*** Lillehammer, 1972
Figs 162, 170, 178, 185, 197.

Nemoura viki Lillehammer, 1972c, Norsk ent. Tidsskr., 19: 161.

Male. Body length 4.2-7.0 mm, wing length 4.7-6.2 mm. Head and thorax dark brown, abdomen and legs light brown to yellow. Pronotum shiny and non-punctate. Middle part of epiproct not overreaching the anterior part (Figs 162, 170, 178). Subgenital plate broad with a small apex, subanal plates varies in form, and may be either uni- or bi-lobed. Cerci with two or three hooks.

Female. Body length 5.0-8.0 mm, wing length 6.2-7.6 mm. Colour similar to male. No structures present beneath the subgenital plate, which is straight or rounded. Segment 7 is unpigmented (Fig. 185).

Nymphs. First instar 0.6-0.7 mm. Fully grown nymphs 5.0-8.0 mm. Body reddish brown. Galea without hairs. Head with almost circular light brown spots. First two antennal segments darker than the rest. Pronotum with a marginal fringe of short bristles (as in Fig. 206). All femora with equally short bristles on posterior margin. Tibiae with both short and long bristles along margins. Abdominal terga 4-6 with a fringe of bristles that may be half as long as tergal length. Apical bristles on cerci more than half as long as length of the individual segments; only a few surface bristles present on the segments (Fig. 197). The nymph was described by Lillehammer (1987d).

Variability. Wide variations occur in the form of the subanal plate, in the form of the hooks on the male cerci, and in the form of the female subgenital plate (Lillehammer, 1974a). Little variation is seen in wing venation.

Distribution. *N. viki* has only been recorded from the northernmost parts of Finland and Norway.

Biology. The species mainly occurs in small slow-flowing streams and in lake outlets where dense scrub of willow (*Salix* spp.) occurs. The habitats seem to represent a northern equivalent to those occupied by *N. dubitans* in the south. The species has a one-year life cycle, adults emerging in June and July. Both the egg development and the nymphal growth is greatly influenced by water temperature (Lillehammer, 1986b).

Genus *Nemurella* Kempny, 1898

Nemura (Nemurella) Kempny, 1898, Verh. zool.-bot. Ges. Wien, 48: 59.
Type species: *Nemoura inconspicua* Kempny, 1898 (mon.).

Small to middle-sized, rufous brown stoneflies, with long light brown wings. Characterized by the strongly modified genital appendages of the male (Fig. 163) and the particular shape of the female subgenital plate which is large and pointed posteriorly (Fig. 186). Nymphs recognizable on the equal length of 1st and 3rd segments of hind tarsus. All legs with a transverse row of long and stout bristles.

29. *Nemurella pictetii* Klapálek, 1900
Figs 163, 171, 186, 190.

Nemurella Pictetii Klapálek, 1900, Rozpr. české Akad. Cís. Fr. Jos., II, 9: 30.
Nemura inconspicua Kempny, 1898, Verh. zool.-bot. Ges. Wien, 48: 49 (preocc.).

Male. Body length 5.4-8.5 mm, wing length 6.1-8.6 mm. Head and thorax dark brown, abdomen rufous brown, wings light brown. Epiproct bilobed, subgenital plate broad with a large forked apex; subanal plates small and simple (Figs 163, 171).

Female. Body length 6.0-11.4 mm, wing length 7.6-10.5 mm. Colour similar to male. Subgenital plate large and more or less triangular, sternum 8 with two large sclerotised structures laterally (Fig. 186).

Nymphs. Fully grown nymphs 7-9 mm. Body dark brown to rufous brown or grey-brown. Head dark brown, ocelli invisible, antenna long and stout. Pronotum with fringe of stout short bristles. Legs with a transverse row of long and stout bristles, the longest bristles being about as long as the femoral width (Fig. 190). Segments 1 and 3 of hind tarsus of about equal length. Cerci long, with strong stout bristles.

Variability. The form of the male genital appendages is fairly stable, while the form of the female subgenital plate varies (Lillehammer, 1974a). The species is nearly always long-winged. Variation in wing venation occurs to the same degree as in most *Nemoura* species. Some variation is also found in the arrangement of the bristles on the legs of the nymphs, the transverse row of stout bristles being only weakly developed in some specimens.

Distribution. *N. pictetii* occurs in all parts of Norway, Finland, Sweden and Denmark, including the islands. The species may have entered Fennoscandia from both the south and the north-east. The species is recorded all over Europe, eastward to Siberia.

Biology. *N. pictetii* occurs in small streams, springs, large rivers, large lakes, and also in smaller water-bodies. The species may have a one-year life cycle under favourable conditions, but several instances of a two-year life cycle are known (Brittain, 1978). Other studies have indicated that the nymphal growth is but little influenced by the water temperature (Elliott, 1984). Emergence in April-September.

Genus *Protonemura* Kempny, 1898

Nemura (Protonemura) Kempny, 1898, Verh. zool.-bot. Ges. Wien, 48: 51.
Type species: *Nemoura meyeri* Pictet, 1841 (des. Illies, 1966).

Small to medium of size, dark to light brown in colour. Males characterised by the shape of the epiproct (Figs 208-210), females by the shape of the subgenital plate (Figs 211-213). Wings light brown with darker veins and held flat over the body. The cubitus area of the forewing with several cross-veins (Fig. 135); often darker bands are present on the wings. Long-winged species.

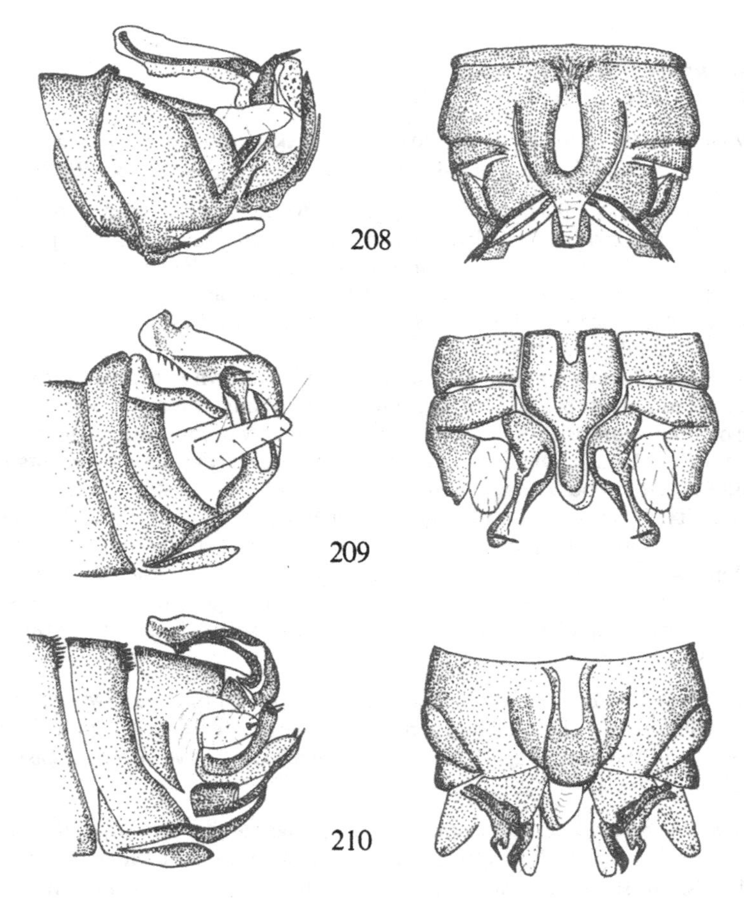

Figs 208-210. Male abdominal apex of *Protonemura*, left row in lateral view, right row in ventral view. - 208: *P. hrabei* Raušer; 209: *P. intricata* (Ris); 210: *P. meyeri* (Pictet).

The nymphs are rufous brown to dark green-grey. The various occurrence of discrete tergites and tergites of the first abdominal segments and the arrangements of the bristles on the cerci and abdominal segments are used to separate the species. The characteristic sausage-like gills are shown in Fig. 139.

Key to species of *Protonemura*

Adults

1 Male: epiproct with simple, straight, dorsal profile-line (Fig. 208). Female: subgenital plate large, covering most of sternum 8 (Fig. 211) .... 30. *hrabei* Raušer

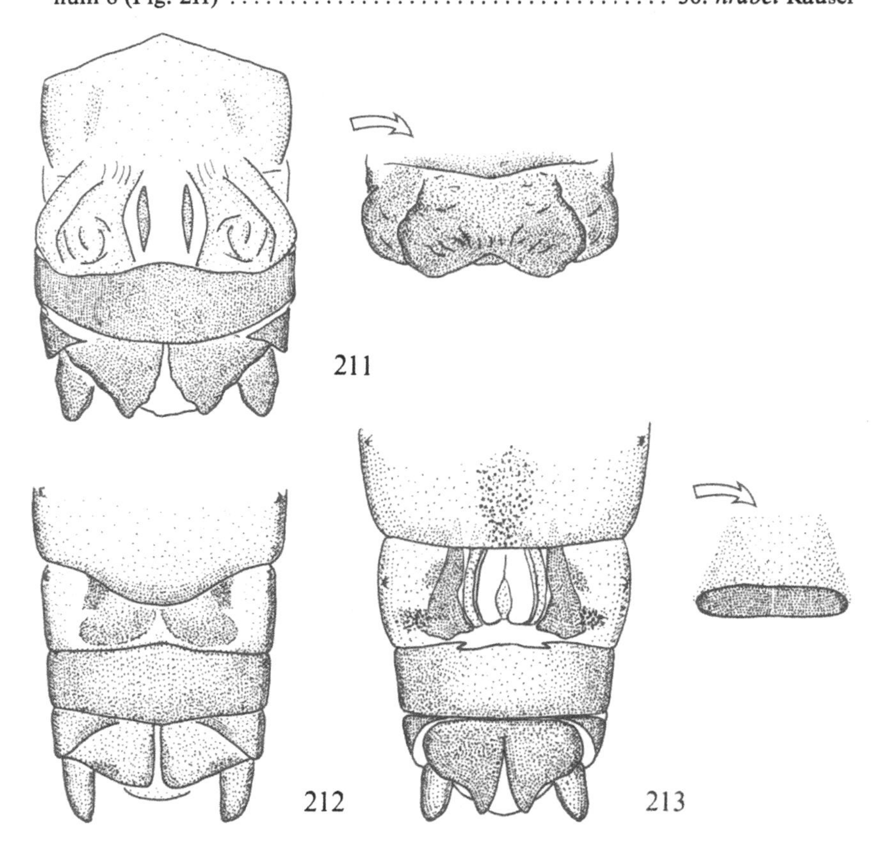

Figs 211-213. Female abdominal apex of *Protonemura,* ventral view. In 211 and 213 the subgenital plate is removed to show the structure beneath the plate. - 211: *P. hrabei* Raušer; 212: *P. intricata* (Ris); 213: *P. meyeri* (Pictet).

- Male: epiproct with a groove dorsally (Figs 209, 210). Female: subgenital plate smaller (Figs 212, 213) 2

2(1) Male: epiproct thick, dorsal margin with deep groove subapically (Fig. 209). Female: subgenital plate divided into two pigmented parts (Fig. 212) .... 31. *intricata* (Ris)

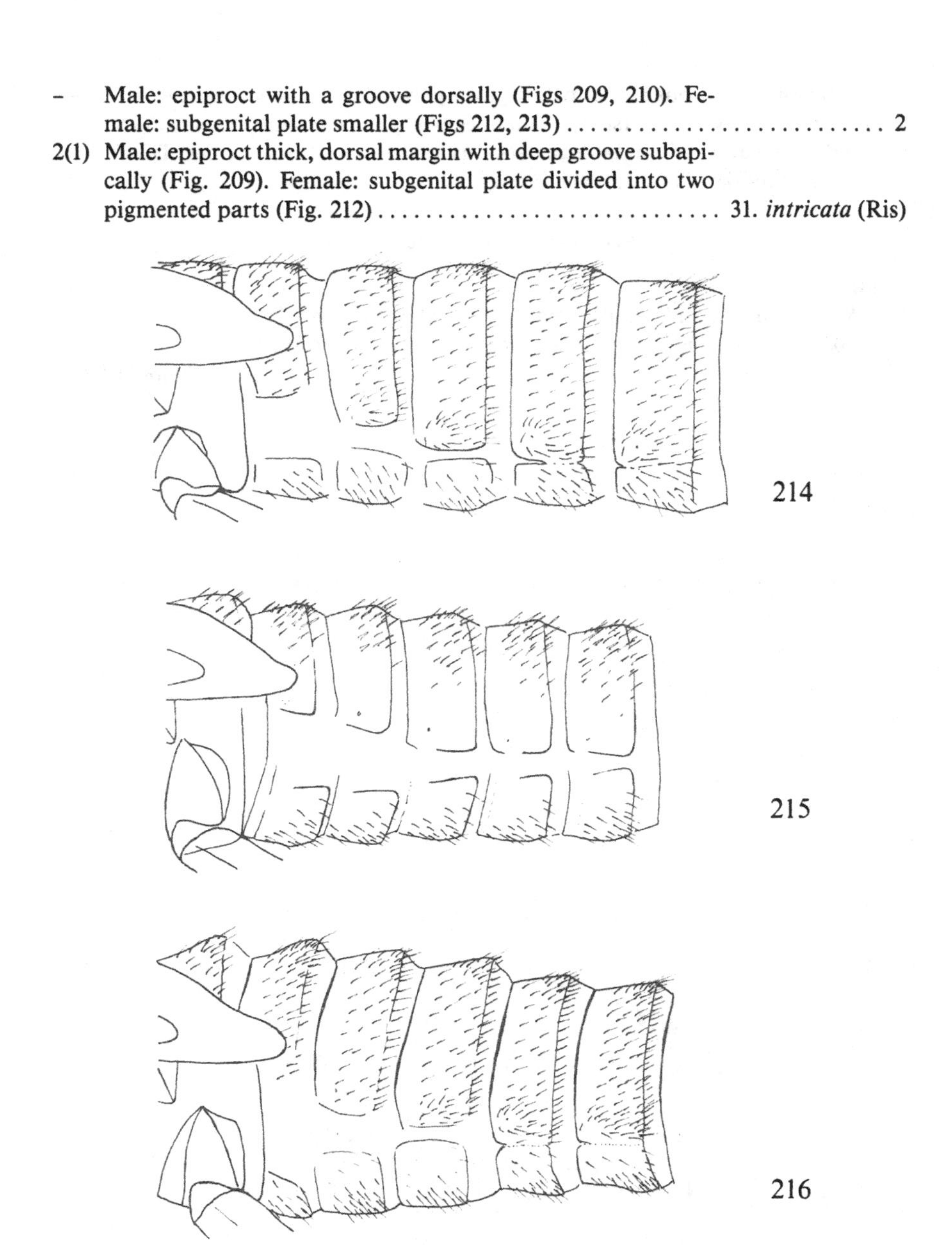

Figs 214-216. Lateral view of abdominal segments 1-6 of *Protonemura* nymphs. - 214: *P. hrabei* Raušer; 215: *P. intricata* (Ris); 216: *P. meyeri* (Pictet).

- Male: epiproct slender, dorsal margin with a shallow groove subapically (Fig. 210). Female: subgenital plate narrowly pigmented in the center of sternum 8 (Fig. 213) .... 32. *meyeri* (Pictet)

Nymphs

1 Abdominal segments 1-5 with discrete tergum and sternum; segments 6-8 partially divided (Fig. 214). Abdominal terga with both long and short bristles (Fig. 217). Cercal segments 7-10 much longer than broad, with a sparse fringe of long and short bristles (Fig. 220) .... 30. *hrabei* Raušer

- Abdominal segments 1-6 with discrete tergum and sternum (Fig. 215). Abdominal terga with both long and short bristles (Fig. 218). Cercal segments 7-10 slightly longer than broad, with a sparse fringe of both long and short bristles (Fig. 221) .... 31. *intricata* (Ris)

- Abdominal segments 1-4 with discrete tergum and sternum; segments 5-8 partially divided (Fig. 216). Abdominal terga with a fringe of bristles of equal length (Fig. 219). Cercal segments 7-10 short, with a dense fringe of short bristles (Fig. 222) .... 32. *meyeri* (Pictet)

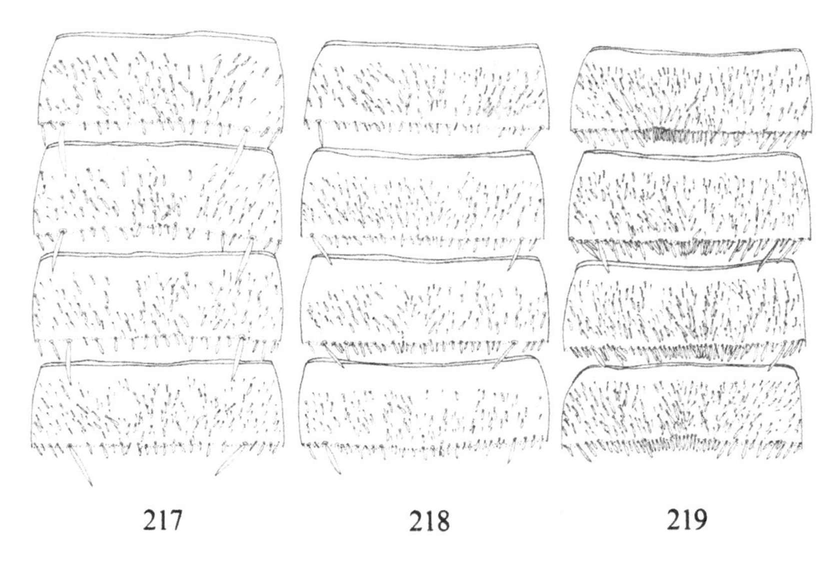

Figs 217-219. Abdominal terga 7-10 of *Protonemura* nymphs. - 217: *P. hrabei* Raušer; 218: *P. intricata* (Ris); 219: *P. meyeri* (Pictet).

30. ***Protonemura hrabei*** Raušer, 1956
Figs 208, 211, 214, 217, 220.

Protonemura hrabei Raušer, 1956, Acta Acad. Sci. Čsl., Brno, 28: 466.

Male. Body length 5.0-8.0 mm, wing length 6.2-10.0 mm. Body chocolate brown, lighter brown on the ventral side of the abdomen; wings light brown with darker veins. The male is always long-winged. Epiproct long and slender, vesicle long and narrow (Fig. 208). Subgenital plate broad and pointed at apex; subanal plates short and stout, outer lobe pointed at apex.

Female. Body length 7-9 mm, wing length 8-11 mm. Colour similar to male. The female is always long-winged. Subgenital plate very large and bi-lobed, covering most of segment 8. Characteristic structures present below the plate (Fig. 211).

Nymphs. Body chocolate brown, abdominal segments 1-5 with discrete tergum and sternum (Fig. 214). Pronotum with a dense row of short bristles. Posterior margin of abdominal terga 7-10 with a few long stout bristles laterally (Fig. 217). Cerci long and slender, relatively few surface bristles present on the cercal segments (Fig. 220).

Variability. Too few specimens were examined for any conclusion to be drawn.

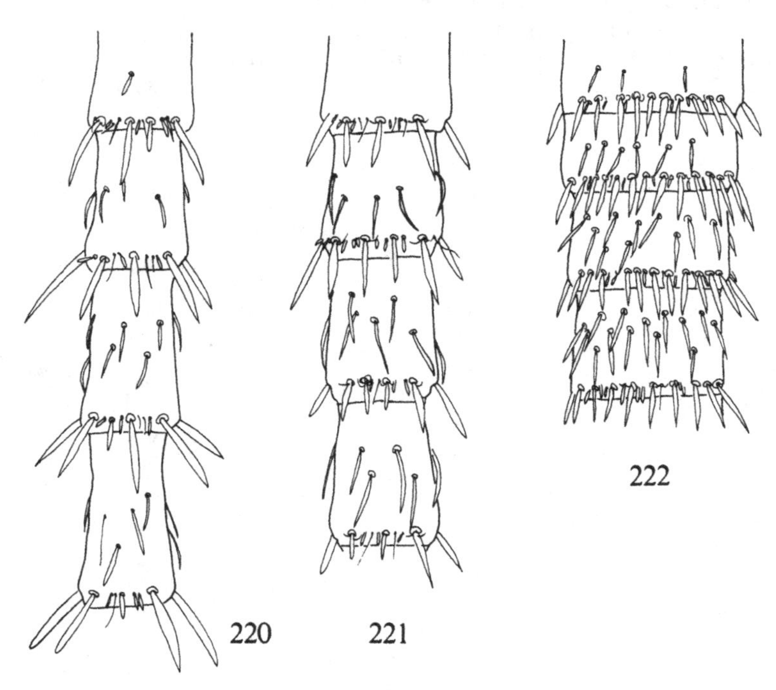

Figs 220-222. Cercal segments 7-10 of *Protonemura* nymphs. – 220: *P. hrabei* Raušer; 221: *P. intricata* (Ris); 222: *P. meyeri* (Pictet).

Distribution. *P. hrabei* has only been recorded at a few localities in Denmark: EJ and NEJ (Jensen *et al.*, 1986), and does not occur in Finland, Sweden or Norway. The species has also been recorded from central Europe but not from Great Britain.

Biology. *P. hrabei* occurs in small woodland streams, adult emerging taking place in September-October. The species may overwinter in the egg stage or as small-sized nymphs. The main nymphal growth period is in May-September.

31. ***Protonemura intricata*** (Ris, 1902)
Figs 19, 209, 212, 215, 218, 221.

Nemoura (Protonemura) intricata Ris, 1902, Mitt. schweiz. ent. Ges., 10: 392.
Nemoura clavata Navás, 1918, Mems R. Acad. Cienc. Artes Barcelona, 14: 347.
Protonemura umbrosa s. Illies, 1966: 245 (nec Aubert, nec A. E. Pictet).
Protonemura umbrosa intricata Consiglio, 1967, Fragm. ent., 5: 37.

Male. Body length 6.3-7.2 mm, wing length 8.0-9.1 mm. Body rufous brown to dark brown, wings light brown with darker veins. Epiproct thick-bodied and with a groove dorsally (Fig. 209). Subgenital plate longer than broad, with very long and pointed apex. Subanal plates bi-lobed, the outer lobe rounded at apex, the inner pointed. Male is long-winged.

Female. Body length 6.7-7.6 mm, wing length 8.3-9.5 mm. Body colour similar to male. Wings smoky grey or brown with dark veins. Posterior margin of sternum 7 rounded. Subgenital plate rounded; on segment 8 two sclerotised lateral structures, both dark pigmented (Fig. 212).

Nymphs. Body rufous brown, darker on the dorsal side. Abdominal segments 1-6 with discrete tergum and sternum (Fig. 215). Posterior margin of abdominal terga 7-10 with only a few bristles laterally (Fig. 218). Pronotum with a row of sparse bristles. Cerci with bristles on the surface of the segments (Fig. 221).

Variability. Too few specimens available for analysis.

Distribution. *P. intricata* has been recorded from both southern and northern parts of Finland, and from the northern parts of the Finnmark province of Norway. The species is absent from the rest of Norway, and from Sweden and Denmark. The species is a south-eastern immigrant to Fennoscandia (Fig. 19).

Biology. The species is recorded from small streams. *P. intricata* has a one-year life cycle, and adults emerge in June and July. The ecology of some European populations has been studied by Zwick (1981) and Brock (1986).

32. ***Protonemura meyeri*** (Pictet, 1841)
Figs 210, 213, 216, 219, 222.

Nemoura meyeri Pictet, 1841, Perlides: 390.
Nemoura subulata Navás, 1917a, Revta. R. Acad. Cienc. Madr., 15: 741.

Nemoura salai Navás, 1927, Boln. Soc. ent. Esp., 10: 82.

Male. Body length 6.8-8.0 mm, wing length 7.9-9.5 mm. Body dark brown-grey, ventral side lighter. Wings light brown with darker veins. Legs long, slender and grey-brown. Segments 8 and 10 with stout bristles on dorsal surface. Epiproct slender, vesicle short, stout and broad (Fig. 210). Subgenital plate as broad as long. Subanal plates bi-lobed, the outer lobe rounded, with pointed inner part.

Female. Body length 7.0-9.9 mm, wing length 8.4-11.4 mm. Colour similar to male. Legs brown. Posterior margin of sternum 7 straight. Subgenital plate small and oval, covering the inner structures that consist of two lobes with a groove in the middle (Fig. 213).

Nymphs. Fully grown nymph 7-10 mm. Body dark brown to dark green-grey. Abdominal segments 1-4 with discrete tergum and sternum (Fig. 216). Along posterior margin of abdominal terga 7-10 nearly evenly long stout bristles (Fig. 219). Pronotum with a fringe of stout bristles. Several stout bristles on the surface of the cercal segments (Fig. 222).

Variability. Marked morphological variations occur in the genitalia of both males and females. In the male this is mainly in the form of the subanal plates and of the tigellus. In the female the widest variation is in the form of the subgenital plate (Lillehammer, 1974a). The length of the wings are fairly constant, but there are some variation in the wing veins. Thorup (1967) mentioned that Danish *P. meyeri* can be separated from *P. hrabei* on the S-curved shape of Rs1, Rs2, M and Cu veins. However, this character shows a wide variation in other Fennoscandian material.

Distribution. *P. meyeri* is in Denmark found in Jutland, but not on the islands. It occurs in all parts of Norway and Sweden, but is distributed mainly over the northern part of Finland. Based on the present records, the species is considered to be a south-western immigrant to Fennoscandia. The species occur over a greater part of Europe including Spain, Italy and Great Britain.

Biology. *P. meyeri* has been recorded from both small and large streams, and also in some lakes in northern Fennoscandia. It has been recorded from sites as high as 1100-1200 m above s.l. and in the low-alpine vegetation belt. The species has a one-year life cycle; adults emerge in March-July.

Family Capniidae

Small, dark-coloured stoneflies with smoky grey wings. Head broader than pronotum, with long antennae and several segments in the cerci (Fig. 37).

The family is characterised by the reduction of the Cu cross-veins, the specialised external copulatory organ of the male, and the absence of a receptaculum seminis in the female.

Nymphs of Capniidae can be distinguished from those of Leuctridae by the

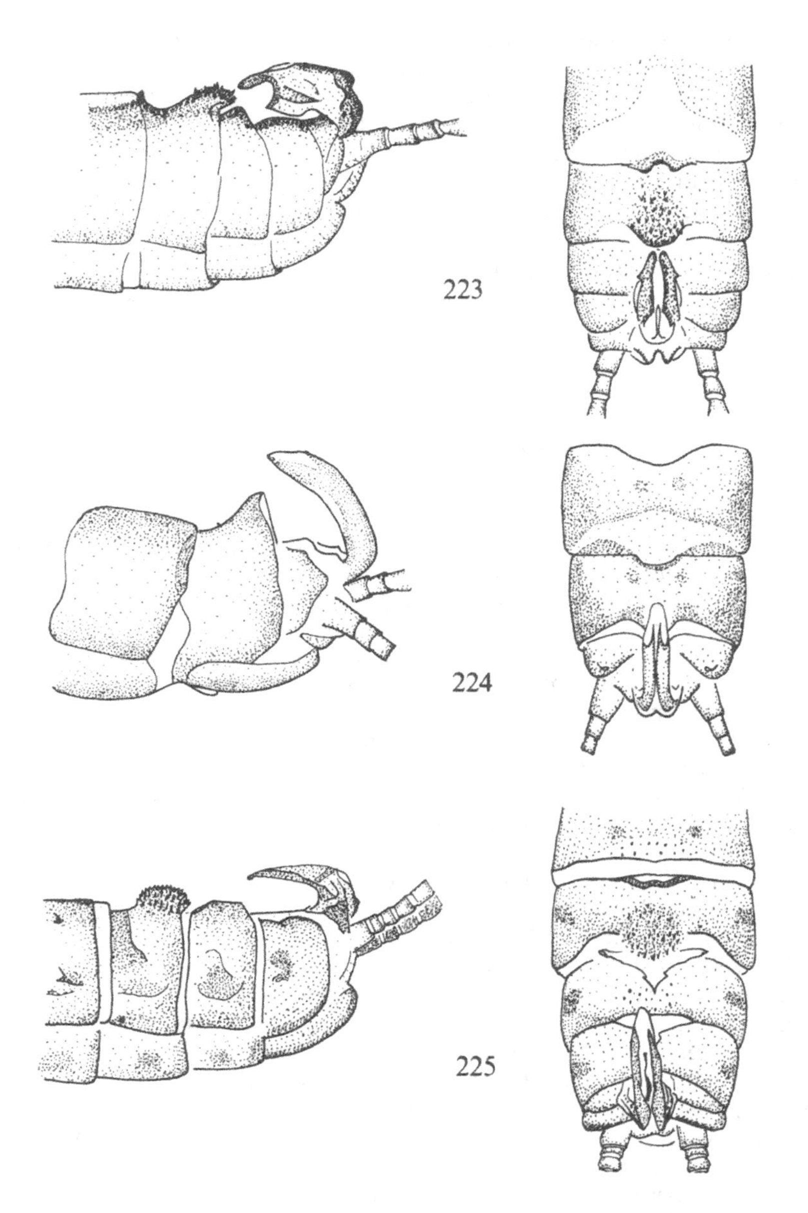

Figs 223-225. Male abdominal apex of *Capnia,* left row in lateral view, right row in dorsal view. - 223: *C. atra* Mort.; 224: *C. bifrons* (Newm.); 225: *C. nigra* (Pict.).

presence of discrete tergal and sternal plates on all the abdominal segments, and in that the cerci of first instar nymphs have three segments, while in Leuctridae they have four segments. The cercal segments are also of different shape (Fig. 51). The nymphs can be separated from nymphs of Nemouridae by the hind legs; when stretched backwards they do not overreach the abdominal apex (Fig. 54).

The family is divided into 14 genera, two of which occur in Fennoscandia and Denmark. Our two genera can be separated from each other by the presence of an anal lobe on the hind wing in *Capnia;* this is absent in *Capnopsis.* In *Capnopsis* the cerci have 7-8 segments, while those of *Capnia* have several additional segments.

Five species of *Capnia* and *Capnopsis schilleri* occur in Fennoscandia and Denmark.

Key to species of Capniidae

Adults

1 Anal area of hind wing reduced, as shown in Fig. 9 .... 38. *Capnopsis schilleri* (Rostock)

– Anal area of hind wing well developed 2

2(1) Male: tergum 9 with a dorsal knob, epiproct knife-shaped (Fig. 224). Female: subgenital plate simple (Fig. 230). Male short-winged, female full-winged .... 34. *Capnia bifrons* (Newman)

– Male: tergum 7 with a dorsal knob. Female: subgenital plate of various shape 3

3(2) Male: epiproct simple and pointed (Fig. 225). Female: central part of subgenital plate strongly sclerotised (Fig. 231). Both sexes full-winged .... 35. *Capnia nigra* (Pictet)

– Male: epiproct distinctly marginate ventrally. Female: subgenital plate different 4

4(3) Male: apex of epiproct deeply incurved (Figs 223, 226). Female: subgenital plate triangular, straight or incurved, with sclerotised part in the middle 5

– Male: epiproct slightly incurved at apex (Fig. 227). Female: subgenital plate triangular without any sclerotised part (Fig. 233) .... 37. *Capnia vidua* Klapálek

5(4) Male: posterior part of epiproct ends in a right angle (see arrow in Fig. 223); tergum 6 with a sclerotised part posteriorly. Female: subgenital plate triangular or rounded, without any sclerotised structure beneath (Fig. 229) .... 33. *Capnia atra* Morton

– Male: posterior part of epiproct rounded (Fig. 226); tergum 6 without sclerotised part posteriorly. Female: subgenital plate triangular or rounded, with two well sclerotised parts beneath (Fig. 232) .... 36. *Capnia pygmaea* Zetterstedt

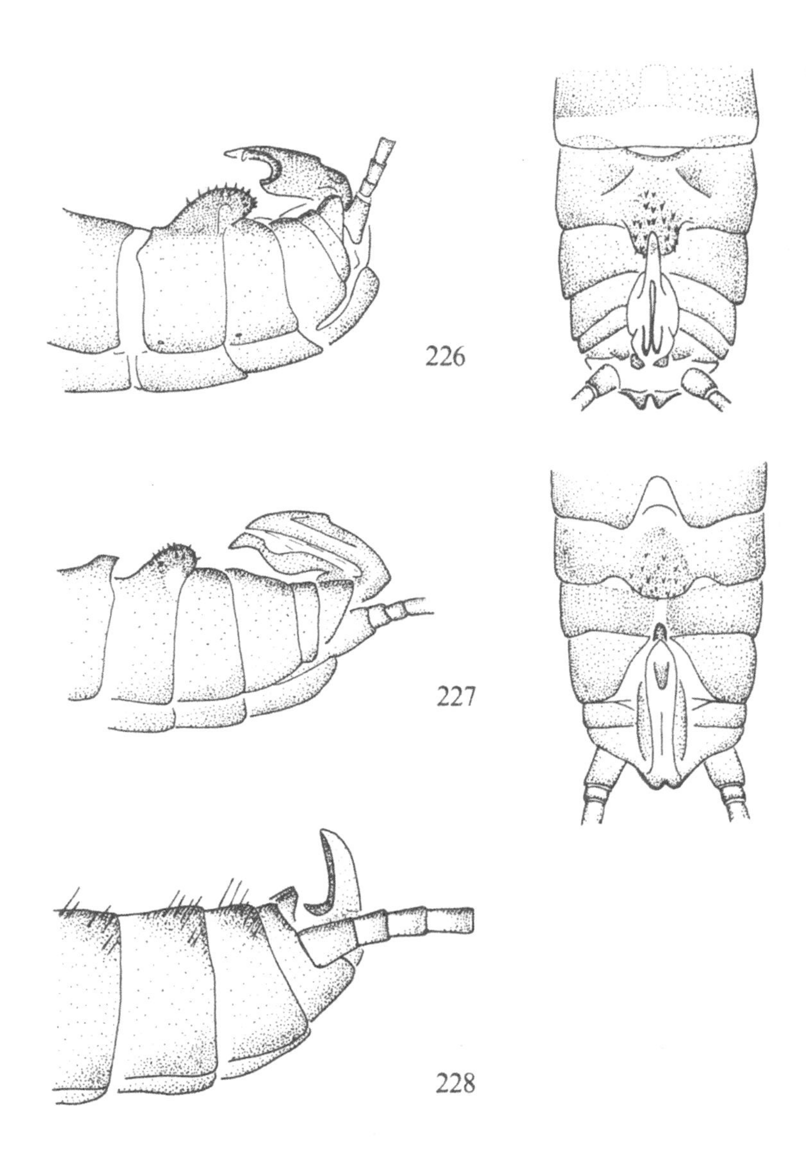

Figs 226-228. Male abdominal apex of *Capnia* and *Capnopsis*, left row in lateral view, right row in dorsal view. - 226: *Capnia pygmaea* (Zett.); 227: *C. vidua* Klap.; 228: *Capnopsis schilleri* (Rostock).

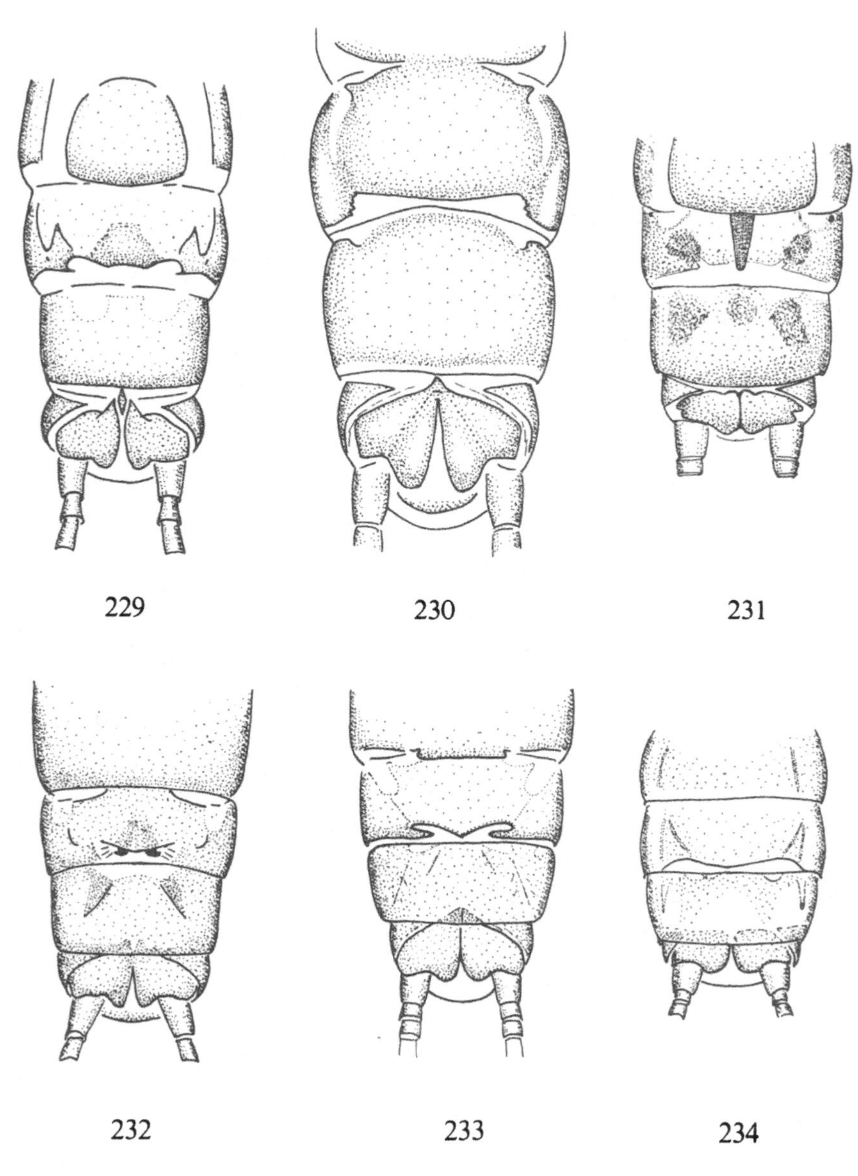

Figs 229-234. Female abdominal apex of *Capnia* and *Capnopsis*, ventral view. – 229: *Capnia atra* Mort.; 230: *C. bifrons* (Newm.); 231: *C. nigra* (Pict.); 232: *C. pygmaea* (Zett.); 233: *C. vidua* Klap.; 234: *Capnopsis schilleri* (Rostock).

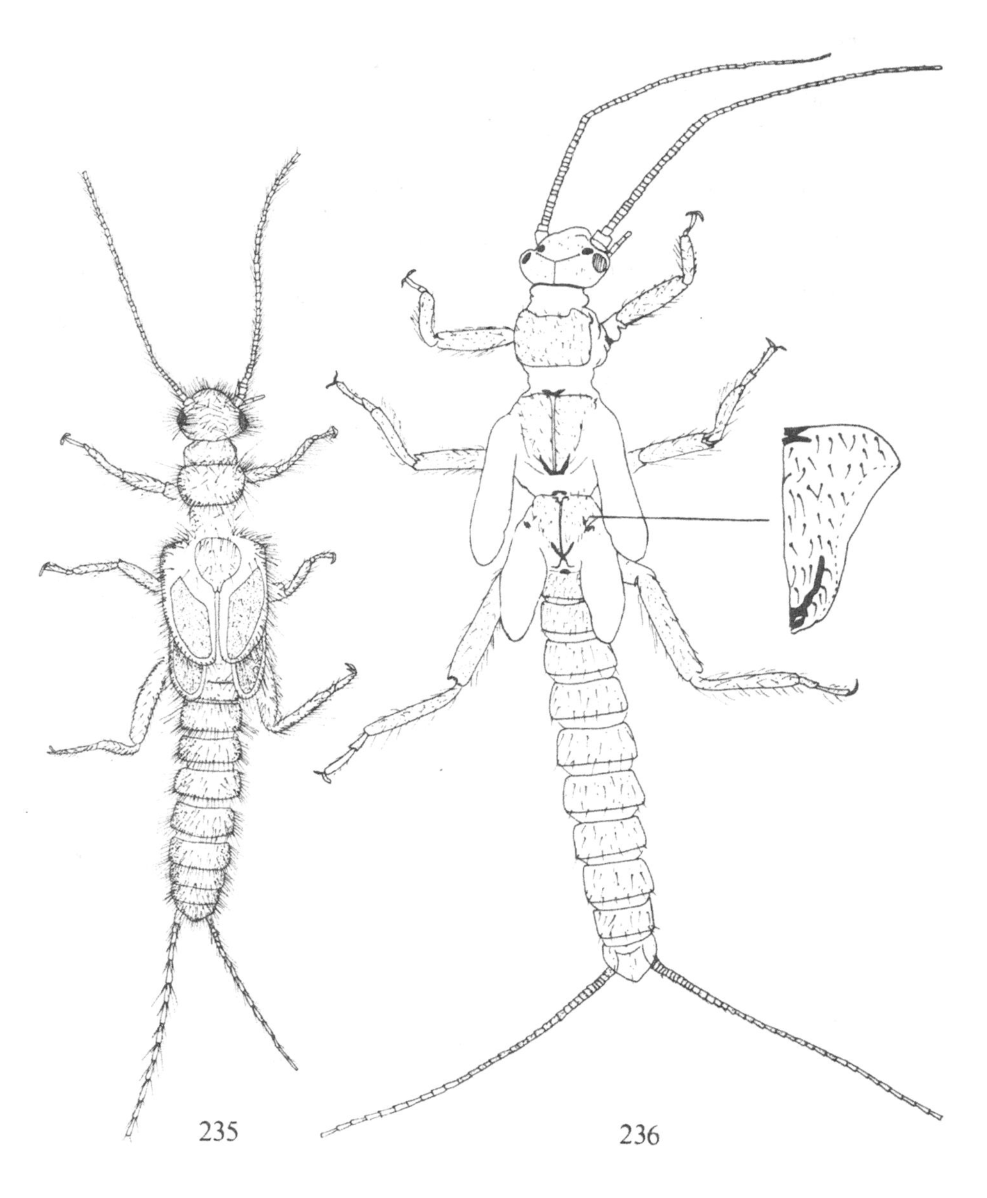

Figs 235, 236. Nymphs in dorsal view of 235: *Capnopsis schilleri* (Rostock) and 236: *Capnia atra* Mort.

Nymphs

1 Whole body covered by long bristles; eyes with a fringe of bristles (Fig. 235) .... 38. *Capnopsis schilleri* (Rostock)
- Whole body covered with short bristles; eyes without a fringe of long bristles (Fig. 236) 2

2(1) Segments of maxillar palp with several fine bristles, terminal segment pointed (Fig. 238) .... 36. *Capnia pygmaea* (Zetterstedt)
- Segments of maxillar palp with few bristles, terminal segment rounded (Fig. 237) 3

3(2) Bristles on cercal segments 7-10 and 15-17 about twice as long as segments (Fig. 240) 4
- Bristles on cercal segments 7-10 and 15-17 less than twice as long as segments (Fig. 241) 5

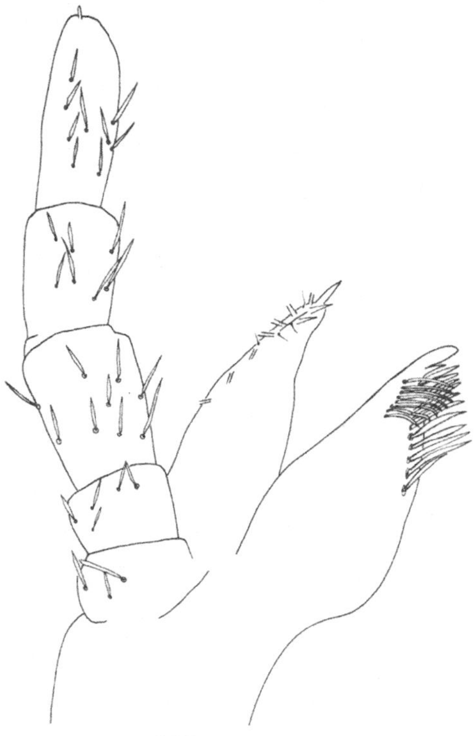

237  238

Figs 237, 238. Nymphal maxilla with palp of 237: *Capnia atra* Mort. and 238: *C. pygmaea* (Zett.).

4(3) Metanotum with irregular rows of distinct bristles (Fig. 236) .... 33. *Capnia atra* Morton
- Metanotum with more regular rows of minute bristles (Fig. 239) .... 35. *Capnia nigra* (Pictet)

5(3) Bristles along dorsal femoral margin long and thin; those along ventral femoral margin short and stout (Fig. 242) .... 34. *Capnia bifrons* (Newman)
- Bristles along dorsal femoral margin of similar length as those along ventral femoral margin (Fig. 243) .... 37. *Capnia vidua* Klapálek

Genus *Capnia* Pictet, 1841

Capnia Pictet, 1841, Perlides: 318.
Type species: *Perla nigra* Pictet, 1833 (des. Enderlein, 1909).

Small to middle-sized stoneflies of dark to blackish colour, and with smoky grey wings. Anal area of hind wing well developed. Head broader than pronotum. Males are characterised by the tergal processes on the 6th to 9th segments and by the distinctive shape of the epiproct (Figs 223-227). Abdominal terga and sterna separated by a membraneous area. Segment 9 is complete, while segment 10 is reduced. Sternite 9 of males form a subgenital plate that may possess a blackish base. Females have a subgenital plate of characteristic shape (Figs 229-233); subanal plates are triangular in form. Terga 1-8 weakly sclerotised in the middle. Cerci are long, with several segments.

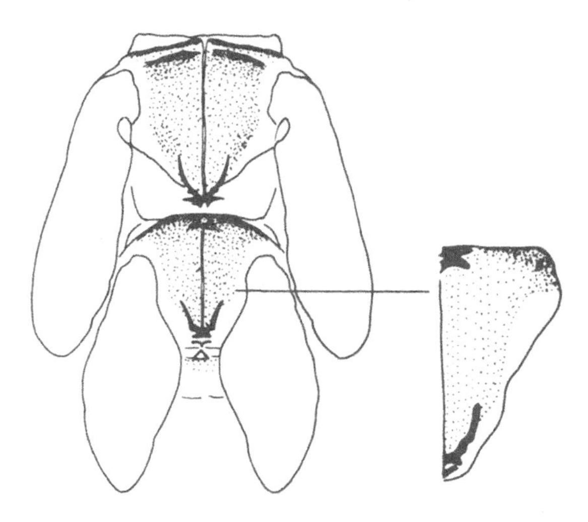

Fig. 239. Meso- and metanotum in dorsal view of nymph of *Capnia nigra* (Pict.).

Nymphs yellow, or light brown to dark brown in colour, and long and cylindrical in shape (Fig 236).

Five species of this genus occur in Fennoscandia and Denmark.

33. ***Capnia atra*** Morton, 1896
Figs 33, 223, 229, 236, 237, 240.

Capnia atra Morton, 1896, Trans. ent. Soc. Lond., 1896: 58.
Capnia praerupta Bengtsson, 1933, Lunds Univ. Årsskr., 29: 34.
Capnia ahngeri Koponen, 1949, Annls Ent. Fenn., 15: 9.

Male. Body length 4.3-6.7 mm, wing length 3.9-6.5 mm. Body black and wings smoky grey with dark veins. Tergite 7 with a dark pigmented dorsal knob with scattered smaller knobs on the surface. Sometimes also segment 6 may possess a knob. Epiproct strongly sclerotised and distinctly sinuously marginate ventrally, straight dorsally and with a short shaft (Fig. 223). Subgenital plate well developed and without a ventral lobe.

Female. Body length 4.5-10.0 mm, wing length 4.2-8.3 mm. Body and wings black, dorsal side of abdomen brown. Subgenital plate may be straight, triangular, or incurved along posterior margin (Fig. 229). No sclerotised structure present below the plate. Subanal plates broader than long.

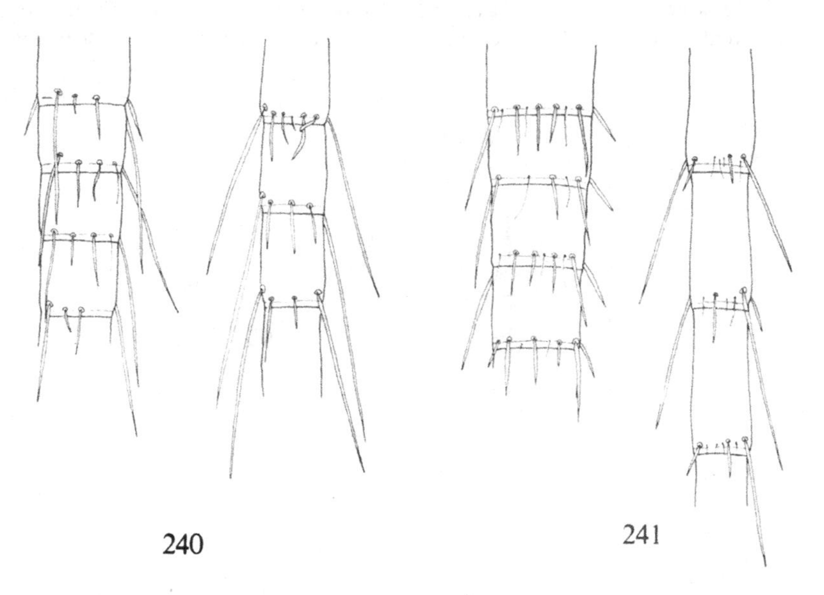

Figs 240, 241. Cercal segments 7-10 and 15-17 of nymphs of 240: *Capnia atra* Mort. and 241: *C. bifrons* (Newm.).

Nymphs. Body long and slender, medium brown on head, pro-, meso-, and metanota, and on the dorsal side of abdomen, light brown ventrally. First instar 0.56-0.70 mm. Fully-grown nymphs 5-10 mm. Terminal segment of maxillar palp blunt-tipped (Fig. 237). Mesonotum with irregular rows of small bristles (Fig. 236). Cerci long and slender with several segments. Bristles on cercal segments 7-10 and 15-17 about twice as long as the respective segmental length (Fig. 240).

Variability. Wing length and the venation pattern are highly variable, and in some

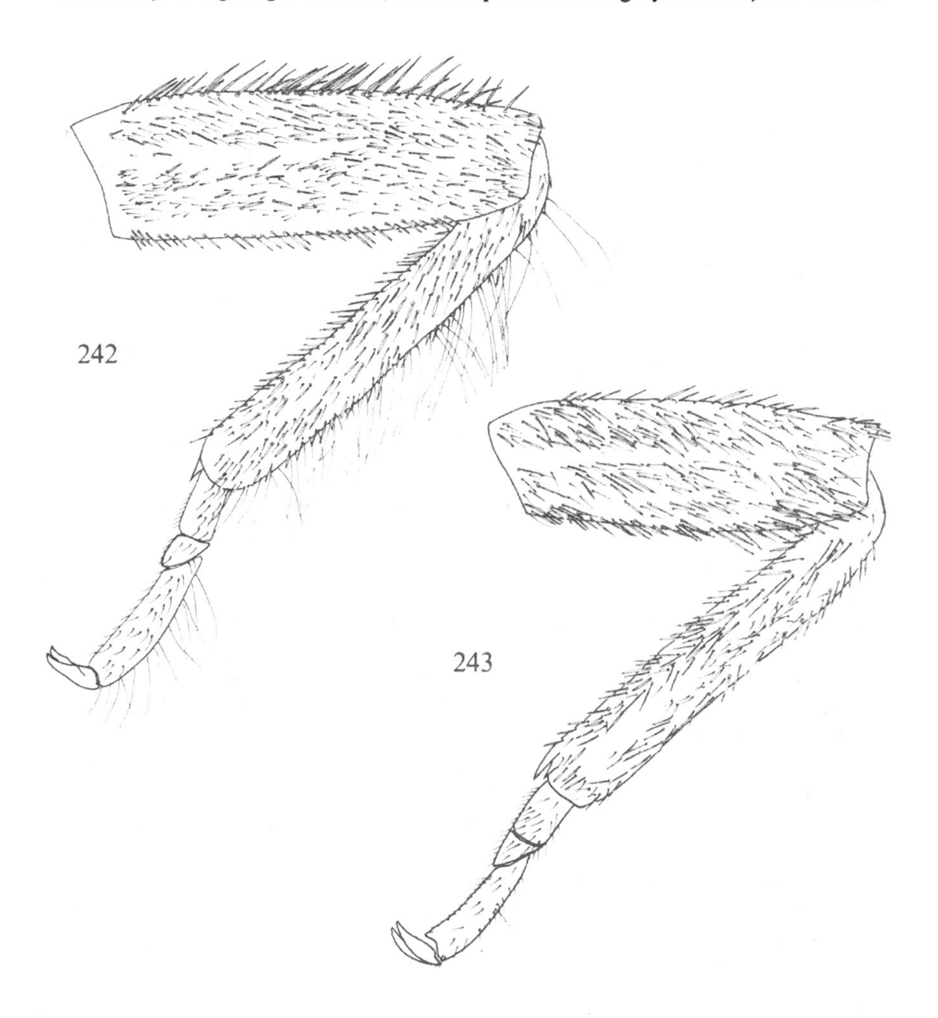

Figs 242, 243. Left hind leg of nymphs of 242: *Capnia bifrons* (Newm.) and 243: *C. vidua* Klap.

populations all individuals have short wings. The form of the genital appendages, such as the tergal structures, shows the widest intraspecific variation in any Capniidae (Lillehammer, 1974a, 1976). Intraspecific variation also occurs in body length of the first nymphal instar. This species is capable of forming local populations that differ both morphologically and ecologically.

Distribution. *C. atra* is most abundant at higher altitudes in Fennoscandia, but occurs also in the lowlands. In Finland the species occurs over a wide area, whereas in southern Sweden it is found at one locality at Lake Vättern. *C. atra* has not been recorded from Denmark. It is considered to be a north-eastern immigrant. Outside Fennoscandia the species only occurs in restricted areas (Fig. 33).

Biology. *C. atra* occurs in small and large streams and is abundant in shallow water in lakes. *C. atra* has a one-year life cycle, adults emerging in March in the lowlands of the south and in June at high altitudes and in the north. The eggs develop rapidly at high water temperatures and slowly at low temperatures (Fig. 14), but nymphal growth (Fig. 15 B) seems to be little influenced by water temperature (Brittain, Lillehammer & Saltveit, 1984, 1986).

34. *Capnia bifrons* (Newman, 1839)

Figs 18, 224, 230, 241, 242.

Chloroperla bifrons Newman, 1839, Ann. Mag. nat. Hist., (2) 3: 89.
Capnia dustmeti Navás, 1917b, Mems R. Acad. Cienc. Artes Barcelona, 13: 6.
Capnia quadrangularis Aubert, 1946, Mitt. schweiz. ent. Ges., 20: 22.

Male. Body length 5.3-8.7 mm, wing length 0.5-0.8 mm. Body dark brown to black, wings lighter brown or smoky grey with darker and strongly reduced veins. Tergum 9 with a dorsal knob. Epiproct knife-shaped (Fig. 224). Subgenital plate broader than long and with a well developed ventral lobe. Males much smaller than females.

Female. Body length 6.9-11.0 mm, wing length 6.8-8.0 mm. Colour similar to male, wings smoky grey with dark veins. The subgenital plate simple (Fig. 230). Subanal plates with a marked inner lobe. This is the largest *Capnia* female.

Nymphs. First instar 0.65-0.75 mm. Fully grown nymphs 6-11 mm. Long and slender, females larger than males. Body red-brown to grey-green dorsally, lighter brown ventrally. Body covered with short bristles. Terminal segment of maxillar palp rounded. Bristles on cercal segments 7-10 and 15-17 less than twice as long as the respective segmental length (Fig. 241). Bristles along dorsal margin of femora longer than those along ventral margin (Fig. 242).

Variability. Both the wing length and the venation pattern are fairly constant, while body length varies strongly in both sexes. Intraspecific variations occur in the shape of the subgenital plate of both sexes. Largest divergence is seen in specimens taken from high altitude localities.

Distribution. *C. bifrons* occurs in Denmark, in south-western Sweden and in eastern Norway. In Norway the species is recorded as far north as the Polar Circle, but does not occur in northern Sweden or in Finland (Fig. 18). The species occurs in parts of Great Britain, in Spain and northern, central, and eastern Europe.

Biology. *C. bifrons* occurs mainly in small streams on a substratum of gravel or small stones. It occurs both in coastal streams and at high altitudes up to 11-1200 m above s.l., above the subalpine vegetation belt. The species has a one-year life cycle, adults emerging in March-June. The species is ovo-viviparous. Nymphal growth is fast at suitable high water temperatures (Lillehammer, 1975b).

35. ***Capnia nigra*** (Pictet, 1833)
Figs 225, 231, 239.

Perla nigra Pictet, 1833, Annls Sci. nat., 28: 61.
Capnia conica Klapálek, 1909b, Čas. české Spol. ent., 6: 101.
Capnia apicalis Navás, 1930, Ark. Zool., 21 (A7): 5.

Male. Body length 4.8-5.5 mm, wing length 5.2-5.9 mm. Body dark brown, wings smoky grey. Tergite 7 with a dark pigmented dorsal knob. Epiproct simple, with a beak-like pointed apex (Fig. 225). Subgenital plate well developed, but without a ventral lobe.

Female. Body length 5.2-5.9 mm, wing length 6.1-7.5 mm. Body colour similar to male. Subgenital plate with a characteristic dark coloured band in the middle (Fig. 231). Subanal plates much broader than long. Segments 8 and 9 with darker pigmented spots laterally. Both sexes are full-winged.

Nymphs. Fully grown nymph 5.5-7.5 mm. Body colour in general yellow-brown dorsally, yellow ventrally. The galea is simple and has no tuft of hairs. Bristles on cercal segments 7-10 and 15-17 about twice as long as the respective segments. Metanotum with rows of minute bristles only (Fig. 239).

Variability. No records on the variability has been made.

Distribution. *C. nigra* has only been recorded from south-eastern Sweden, and is not found in Finland, Norway and Denmark. The species is recorded in southern, central, and eastern Europe.

Biology. Little is known of the biology of this species which occurs in small streams. It is thought to have a one-year life cycle; adult emerging takes place in March-May. Egg development takes 21 days at 15°C (Brinck, 1949).

36. ***Capnia pygmaea*** (Zetterstedt, 1840)
Figs 32, 226, 232, 238.

Perla pygmaea Zetterstedt, 1840, Ins. Lapp.: 1059.
Capnia sparre-schneideri Esben-Petersen, 1910a, Tromsø Mus. Årsskr., 31/32: 83.
Capnia tenuis Bengtsson, 1931, K. svenska VetenskAkad. Skr. Naturskydd., 18: 59.

Male. Body length 3.7-5.7 mm, wing length 4.7-6.2 mm; the smallest of the Fennoscandian *Capnia* species. Body black and wings smoky grey. Tergite 7 with a dorsal knob. Epiproct with a long shaft and a deep incurvation apically; and a deep dorsal insertion (Fig. 226). Colour of epiproct often light yellow. Subgenital plate well developed, without a ventral lobe. Sometimes also tergite 6 may possess a knob.

Female. Body length 4.0-7.5 mm, wing length 5.7-8.3 mm. Body black, wings smoky grey, abdomen light brown dorsally. Subgenital plate triangular or rounded, with two sclerotised structures beneath (Fig. 232).

Nymphs. Long and slender. Body yellow-brown dorsally, yellow ventrally. Body length of the first instar not recorded. Fully-grown nymph 4-8 mm. Galea with some long hairs and terminal segment of maxillar palp pointed (Fig. 238). Bristles on cercal segments 7-10 and 15-17 about twice as long as the respective segments (as in Fig. 240).

Variability. Both wing length and venation pattern are fairly constant in both sexes. *C. pygmaea* is always long-winged. The male genital appendages may vary slightly in shape, but the shape of the female subgenital plate shows a wider variation; especially the two sclerotised structures beneath vary widely in shape.

Distribution. *C. pygmaea* is most abundant in northern Fennoscandia, being largely restricted to the northern part of Finland. It does not occur south of the province of Dalarne in Sweden, nor is it found in the southernmost parts of Norway. The species is also absent from Denmark. *C. pygmaea* is a north-eastern immigrant that does not occur south of Scandinavia (Fig. 32).

Biology. *C. pygmaea* occurs mainly in large rivers in the south but may occur in both small and large rivers in the north. The species has a one-year life cycle, adults emerging in May-June. Nymphal growth mainly takes place at low temperatures during the winter (Baekken, 1981).

37. ***Capnia vidua*** Klapálek, 1904
Figs 30, 227, 233, 243.

Capnia vidua Klapálek, 1904c, Rozpr. české Akad. Cís. Fr. Jos., II, 13: 723.

Male. Body length 4.5-5.5 mm. The male is micropterous, and the venation is irregular, wing length about 1 mm. Body dark brown, wings smoky grey with darker veins. Both segments 6 and 7 with dorsal knobs; on segment 6 the knob is split into two lobes with an insertion in the middle, on segment 7 the knob is rounded and dark pigmented. Apex of epiproct slightly incurved (Fig. 227). The subgenital plate is well developed and without a ventral lobe.

Female. Body length 5.8-7.7 mm, wing length 4.2-6.0 mm. Body brown dorsally, light brown ventrally, wings smoky grey. Wings vary greatly in length, from branchypterous to full-winged. Subgenital plate triangular (Fig. 233), without sclerotised structures beneath.

Nymphs. Medium-brown dorsally, yellow-brown ventrally. Galea without a distinct tuft of hairs. Terminal segment of maxillar palp rounded. Bristles on cercal segments

7-10 and 15-17 less than twice as long as the respective segments. Bristles along dorsal margin of femora about as long as bristles along ventral margin (Fig. 243). The first instar is not known.

Variability. In the Fennoscandian populations wide variations occur both in the shape of male genital appendages, in the shape of the female subgenital plate, and in the wing venation (Lillehammer, 1972, 1974a).

Distribution. *C. vidua* occurs in all the Fennoscandian countries, but only north of the Polar Circle. The species is absent from Denmark, and the nearest populations occur in Iceland and Scotland (Fig. 30). The species consists of several subspecies.

Biology. The species occurs mainly in small streams and is supposed to have a one-year life cycle in Fennoscandia, adults emerging in June. In Iceland this species emerges from March to August and has an unsyncronized growth.

Genus *Capnopsis* Morton, 1896

Capnodes Rostock, 1892, Berl. ent. Z., 37: 3 (preocc.).
Capnopsis Morton, 1896, Trans. ent. Soc. London, 1896: 61.
Type species: *Capnodes schilleri* Rostock, 1892 (mon.).

Small, slender, dark-coloured stoneflies with smoky wings held flat over the abdomen. The genus is characterized by the absence of an anal lobe on the hind wing and the reduced cerci. Labial palps very long. Sternite 9 of male forms a subgenital plate. Nymphs are densely covered by long fine bristles all over the body, antennae and cerci. Only one species occurs in Fennoscandia.

38. *Capnopsis schilleri* (Rostock, 1892)
Figs 228, 234, 235.

Capnodes schilleri Rostock, 1892, Berl. ent. Z., 37: 1.

Male. Body length 3.0-5.2 mm, wing length 5.2-5.8 mm. Body black, wings smoky grey with darker veins. Head broader than pronotum. Epiproct simple (Fig. 228). No knobs on tergites 7 to 9. Subgenital plate strongly sclerotised. Subanal plates fused in the middle.

Female. Body length 3.8-6.0 mm, wing length 5.7-6.9 mm. Colour similar to male. Both sexes are full-winged. The subgenital plate weakly developed (Fig. 234).

Nymphs. First instar 0.65-0.75 mm. Fully grown nymphs 4-6 mm. Long and slender; when stretched backwards the hind leg does not overreach the abdomen. Body dark red-brown dorsally, lighter ventrally. The entire body covered by long fine hairs, likewise the antennae and cerci (Fig. 235). This makes the nymphs easy to separate from those of *Capnia*. The antennae of the first instar has 9 segments, the cerci have 3 segments.

Variability. The wing venation is fairly constant, likewise the form of the genital appendages of both sexes. Variations occur in body length of both males, females and nymphs.

Distribution. The species occur in both the south and the north of Finland, Sweden and Norway, but is absent from south-west Norway and all Denmark. The species may have entered Fennoscandia from the north-east, or from the south, or from both directions. The species is distributed over a larger part of Europe, including Italy. The species occurs in northern Africa, but not in Great Britain.

Biology. *C. schilleri* nymphs occur in both small and large streams, but seem to prefer small streams with a sandy bed. *C. schilleri* has a one-year life cycle, adults emerging in May-July. The eggs have no diapause and at 20°C take only 9-11 days to hatch (Haaland, 1981).

Family Leuctridae

Small, dark insects with wings held wrapped around the abdomen. Cerci simple and one-segmented in both sexes. Terga of males with characteristic structures. Female subgenital plate well developed, often with two lobes.

The family is characterized by the following synapomorphic characters: the internal part of the paraproct is tubular, fused basally and expanded apically to form a cluster; a marked expansion of the vasa deferentia, which functions as an extra sperm sac (Zwick 1973a).

Nymphs long and slender, light in colour and with long cerci composed of several segments. Legs short with marked covering of bristles.

The family is divided into two subfamilies: the Megaleuctrinae Zwick and the Leuctrinae Klapálek. Only the latter subfamily occurs in Europe. Four *Leuctra* species occur in Fennoscandia and Denmark.

Genus *Leuctra* Stephens, 1836

Nemoura (Leuctra) Stephens, 1836, Illus. Brit. Ent., 6: 144.

Type species: *Phryganea fusca* Linnaeus, 1758 (des. Opinion ICZN 836, 1967).

Small, slender, dark pigmented stoneflies. Males characterised by the tergal processes on segments 6-9 (Figs 245-248); the females by possessing a bilobed subgenital plate and a spermatheca (Figs 249-252). Cerci of both sexes with only a single segment (Fig. 38). Wings dark and held wrapped around the body. The cubitus area of the forewing has several cross-veins (Fig. 244).

The nymphs are light coloured, long and slender (Fig. 54). Glossa well developed. Legs short, when stretched backwards not over-reaching the apex of the body. Abdominal segments 1-4 with discrete tergal and sternal plate; 5-9 fused and forming complete rings; sternum 10 reduced (Fig. 48). Subanal plates longer than broad. The first instar nymph has four cercal segments, and an antenna of nine segments.

Key to species of *Leuctra*

Adults

1 Male: processes present on tergum 8 only (Fig. 247). Female: subgenital plate bilobed, each lobe backwardly projecting and rounded apically (Fig. 251) .... 41. *hippopus* Kempny

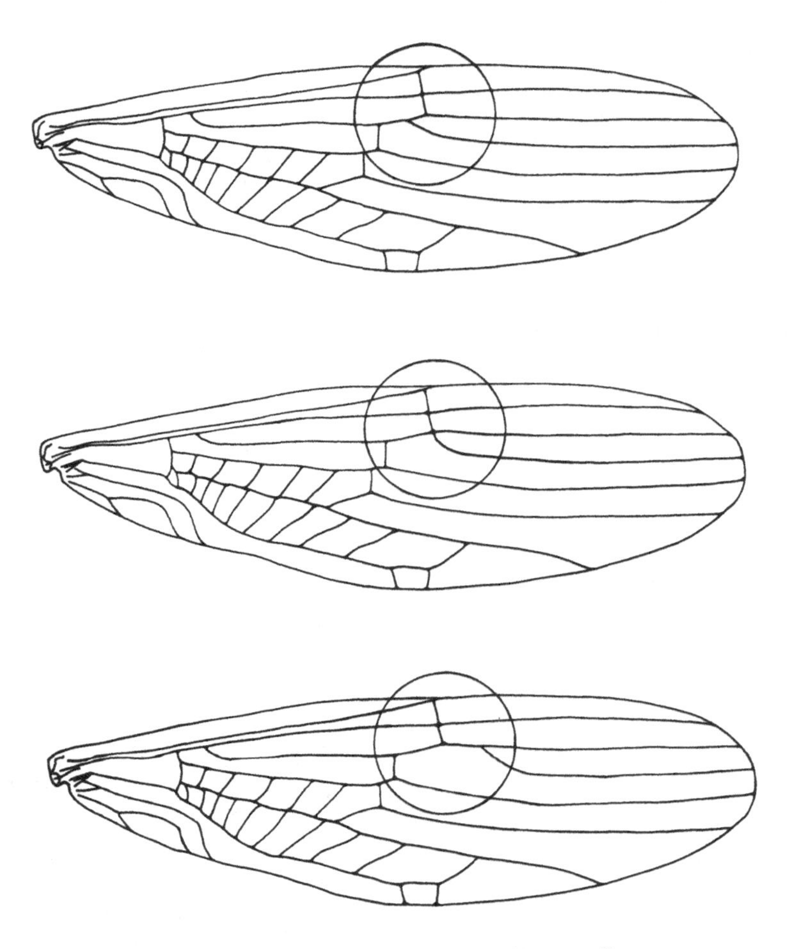

Fig. 244. Variation in venation of *Leuctra hippopus* Kempny.

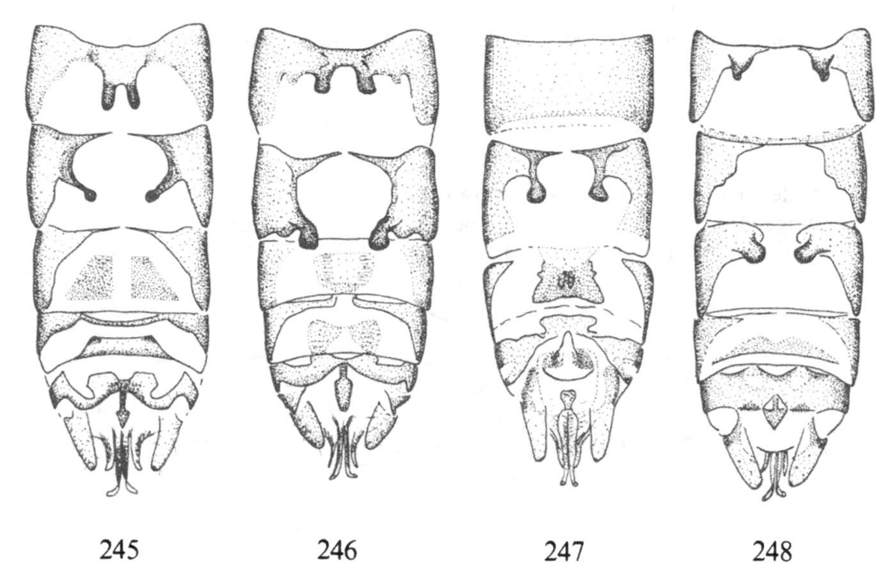

Figs 245-248. Abdominal terga 6-10 of *Leuctra* males, showing tergal processes. – 245: *L. digitata* Kempny; 246: *L. fusca* (L.); 247: *L. hippopus* Kempny; 248: *L. nigra* (Oliv.).

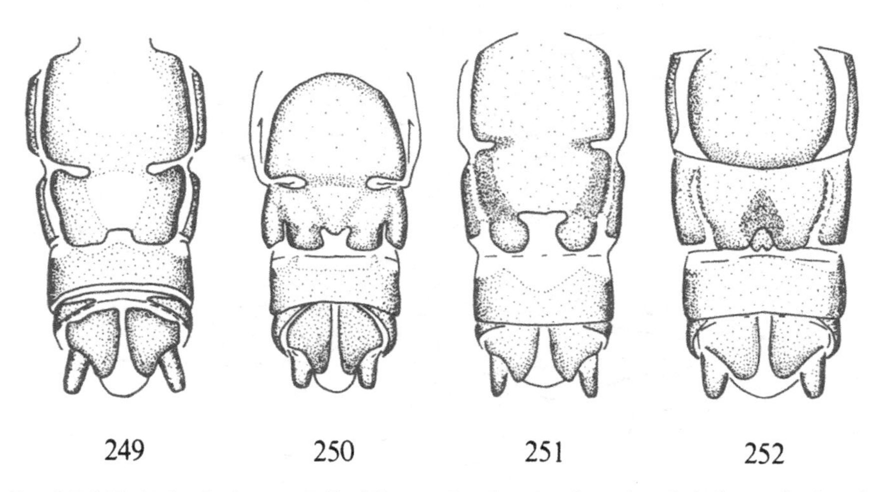

Figs 249-252. Abdominal sterna 7-10 of *Leuctra* females, showing subgenital plate and subanal plates. – 249: *L. digitata* Kempny; 250: *L. fusca* (L.); 251: *L. hippopus* Kempny; 252: *L. nigra* (Oliv.).

– Male tergal processess and female subgenital plate not as described 2

2(1) Male: processes present on terga 6 and 8 (Fig. 248). Female: subgenital plate with a small central lobe between the two main lobes (Fig. 252) . . . . 42. *nigra* (Olivier)

– Male: processes present on terga 6 and 7. Female: subgenital plate without central lobe 4

3(2) Male: tergal processes on segment 6 with two well separated lobes (Fig. 246). Female: subgenital plate bilobed, inner margin incurved (Fig. 250) . . . . 40. *fusca* (Linnaeus)

– Male: tergal processes on segment 6 with two lobes fused at base (Fig. 245). Female: subgenital plate bilobed, inner margin straight (Fig. 249) . . . . 39. *digitata* Kempny

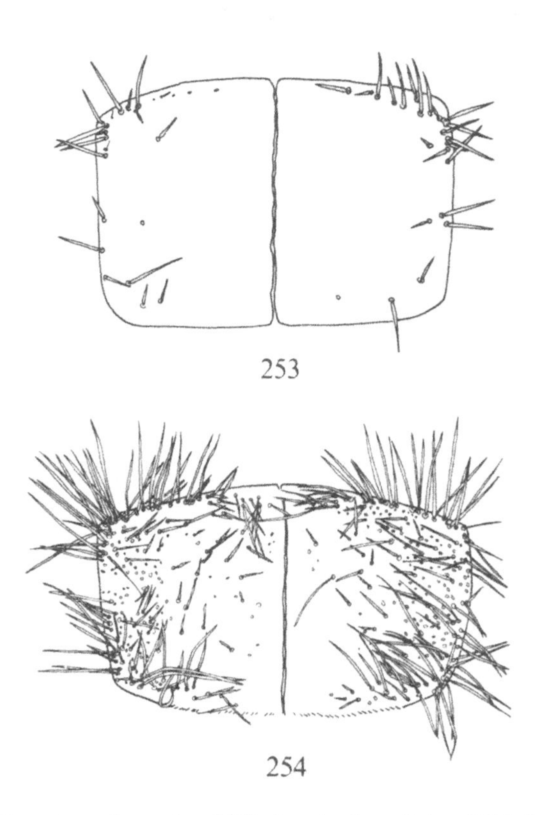

Figs 253, 254. Pronotum of nymphs of 253: *Leuctra fusca* (L.) and 254: *L. nigra* (Oliv.).

Nymphs

1 Pronotum with a dense fringe of long fine hairs (Fig. 254); most of body and legs covered with similar hairs (Fig. 257) .... 42. *nigra* (Olivier)

\- Pronotum with relatively few bristles laterally (Fig. 253); few bristles on the rest of the body 2

2(1) All tibiae with a fringe of long fine hairs (Fig. 255). Abdominal segments 4-7 dorsally with a few long hairs among the short ones (Fig. 258) .... 40. *fusca* (Linnaeus)

\- All tibiae with only a few hairs (Fig. 256) 3

3(2) Abdominal segments 4-7 with short hairs only (Fig. 260). Bristles on cercal segments 7-10 at most half as long as the segmental length (Fig. 262) .... 41. *hippopus* Kempny

\- Abdominal segments 4-7 with both long and short hairs (Fig. 259). Bristles on cercal segments 7-10 at least 3/4 as long as the segmental length (Fig. 261) .... 39. *digitata* Kempny

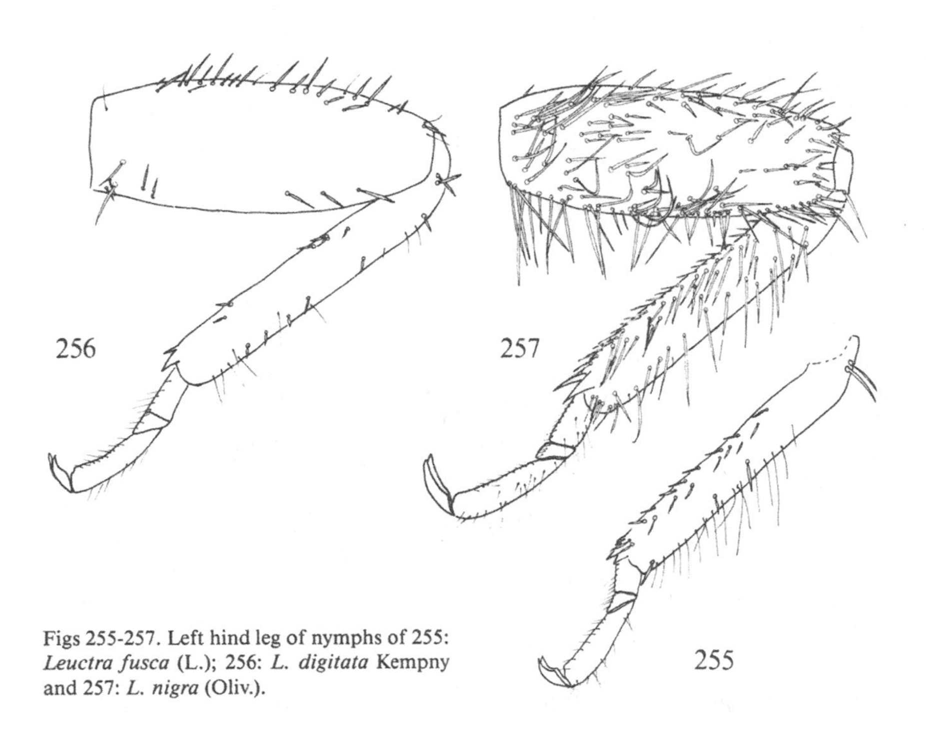

Figs 255-257. Left hind leg of nymphs of 255: *Leuctra fusca* (L.); 256: *L. digitata* Kempny and 257: *L. nigra* (Oliv.).

39. ***Leuctra digitata*** Kempny, 1899
Figs 25, 245, 249, 256, 259, 261.

Leuctra digitata Kempny, 1899, Verh. zool.-bot. Ges. Wien., 49: 13.

Male. Body length 5.0-7.3 mm, wing length 6.8-8.3 mm. Body and wings black, legs and antennae brown. The bilobed tergal processes of segment 6 fused at base. Tergal processes of segment 7 thin, arising about midway between anterior and posterior margin of segment (Fig. 245). Tergal processes of segment 8 forms two plates which sometimes are triangular and sometimes more rectangular in shape.

Female. Body length 5.0-9.0 mm, wing length 6.9-9.1 mm. Body and wings black, tergal part of abdomen unpigmented, antennae and legs brown. The outer and inner margins of the two lobes of the subgenital plate straight (Fig. 249), the middle part often weakly sclerotised.

Nymphs. First instar 0.6-0.7 mm. Fully grown nymphs 5.0-9.0 mm. Body colour yellow-brown to green-brown, darker on the dorsal side. Pronotum with a few bristles laterally. Abdominal terga 4-7 with mainly short bristles, a few long ones scattered over the segmental surface (Fig. 259). Femur and tibia with a few bristles (Fig. 256). Bristles on cercal segments 5-7 of a length equal to the length of the respective segments (Fig. 261). Those on cercal segments 15-17 less than half as long as the respective segments.

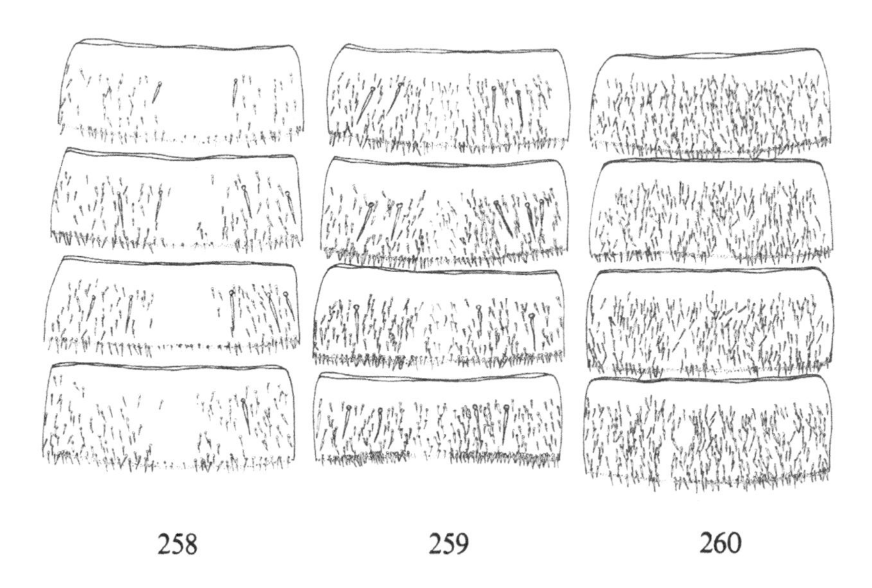

Figs 258-260. Abdominal terga 4-7 of *Leuctra* nymphs. - 258: *L. fusca* (L.); 259: *L. digitata* Kempny and 260: *L. hippopus* Kempny.

Variability. Wing venation fairly constant in both sexes. Marked variation occurs in the tergal structure of segments 6-9 of the male. Some variation is also seen in the shape of the two lobes of the female subgenital plate (Lillehammer, 1974a). The species is always long-winged. Some variation may be found in the bristle covering of the femora and tibiae of the nymphs. Young nymphs may have weakly developed characters and are therefore difficult to separate from nymphs of *L. fusca.*

Distribution. *L. digitata* occurs over a wide area of Fennoscandia, but is absent from the coastal areas of south-western Norway, rare in western Denmark, and absent from southern Sweden (Fig. 25). The species is more often distributed at high altitudes than *L. fusca,* and is more common in the northernmost parts of Fennoscandia. Occurs in all parts of Finland and is probably a north-eastern immigrant to Fennoscandia. The species occurs in central Europe, but has not been recorded in Great Britain.

Biology. The nymph feeds on plant fragments and other detritus. *L. digitata* nymphs occur both in large and small streams, mainly on stony substratum. The species has a one-year life cycle. The main growth period is during the summer months of June-August, adult emerging taking place in July-September. Eggs are laid in August-September. If the nymphs hatch in autumn, the growth is slow during the winter. The life cycle has been studied by Müller & Mendl (1980) and Lillehammer (1985b).

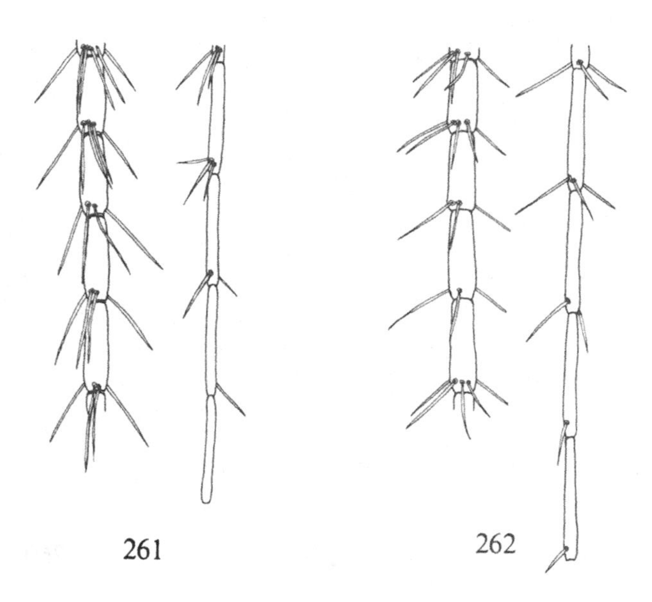

Figs 261, 262. Cercal segments 7-10 and 17-20 of *Leuctra* nymphs. - 261: *L. digitata* Kempny and 262: *L. hippopus* Kempny.

40. ***Leuctra fusca*** (Linnaeus, 1758)
Figs 246, 250, 253, 255, 258.

Phryganea fusca Linnaeus, 1758, Syst. Nat., 10. Ed., 1: 549.
Perla cylindrica De Geer, 1778, Mem. Hist. Ins., 7: 59.
Leuctra fusciventris Stephens, 1836, Illus. Brit. Ent., 6: 145.

Male. Body length 4.4-6.8 mm, wing length 6.2-7.8 mm. Body and wings black, shiny and well sclerotised; antennae and legs brown; wings smoky grey with darker veins. Tergal processes of segment 6 with two separate lobes (Fig. 246); tergal processes of segment 7 broad and arising close to posterior margin of tergite. Always full-winged.

Female. Body length 4.2-8.5 mm, wing length 6.8-9.3 mm. Body black to dark brown, wings grey. Dorsal side of abdomen membraneous, light brown or yellow. Inner side of the two lobes of the subgenital plate curved inwards (Fig. 250); sometimes also the outer side incurved. The middle part of subgenital plate always well sclerotised.

Nymphs. First instar 0.6-0.7 mm. Fully grown nymph 5-9 mm. Both thorax and abdomen light brown. Pronotum with a few bristles laterally. Legs short and tibiae with a fringe of long bristles (Fig. 255). Abdominal terga 4-7 with mainly short bristles (Fig. 258). Bristles on cercal segments 7-10 as long as length of the respective segments, on segments 17-20 less than half as long as the respective segmental length.

Variability. Wing venation of both sexes fairly constant. Wide variations occurs in the shape of the tergal structures of segments 6-9 of the male. Some variation is seen in the shape of the lobes of the female subgenital plate (Lillehammer, 1974a), and likewise in the fringe of long, fine bristles on tibia of the nymphs. They may therefore be difficult to separate from *L. digitata.*

Distribution. *L. fusca* is distributed over most of Fennoscandia, but the species is rare in the northernmost part of the Finnmark province of Norway. The species may have reached Fennoscandia from both the south and the north-east. The species occurs all over Europe, eastward to Siberia.

Biology. The nymph feeds on plant fragments and other detritus. *L. fusca* nymphs occur in both small and large streams as well as in lakes. The species has a one-year life cycle, adults emerging in late summer and autumn. The main growth period is June-August. In some areas emergence takes place over a long period of time. For example, in Øvre Heimdalen, in the mountains of south Norway, emergence occurs from the middle of July to the first half of October (Lillehammer, 1975, 1984). The life cycle has been studied by Brinck (1949), Svensson (1966), and Lillehammer (1985b).

41. ***Leuctra hippopus*** Kempny, 1899
Figs 244, 247, 251, 260, 262.

Leuctra Hippopus Kempny, 1899, Verh. zool.-bot. Ges. Wien, 49: 10.

Male. Body length 4.0-7.1 mm, wing length 5.3-6.9 mm. Body, legs and antennae black, wings smoky grey, abdomen dark pigmented with marked tergal structures. The tergal processes of segment 8 with two separate lobes (Fig. 247). A sclerotised plate present on segment 9, and a supra-anal plate on segment 10.

Female. Body length 4.5-8.8 mm, wing length 5.0-8.2 mm. Colour similar to male. The two lobes of the female subgenital plate are smaller and more rounded than those of other *Leuctra* species (Fig. 251).

Nymphs. First instar 0.70-1.00 mm. Fully grown nymphs 5-9 mm. Both head, thorax and abdomen light brown to yellow in colour. Legs short and stout. Pronotum with a few bristles laterally. Abdominal segments with short bristles only (Fig. 260). Femur and tibia with a few bristles. Bristles on cercal segments 5-7 and 15-17 shorter than segmental length (Fig. 262).

Variability. Wide variations occur in both wing length, venation (Fig. 244), and in the adult body length. Short-winged specimens occur in most populations, and in some all the specimens are short-winged. Wide variations are seen in the form of the tergal processes of the male, and some local populations may differ distinctly from others. Some variation may also be seen in the shape of the microstructures of the sclerotised parts of the abdominal terga (Lillehammer, 1986a). Some variation also occurs in the form of the two lobes of the female subgenital plate, although not to the same extent as seen in the male genitalia. The mean body length of newly hatched nymphs may be significantly different from one population to another. Some differences also occur in other morphometric characters of the nymphs.

Distribution. *L. hippopus* occurs in all parts of Fennoscandia and Denmark, including many of the Danish islands, such as Zealand and Bornholm. However, it is absent from Gotland and Öland, and also from the south-western part of Finland. The species is recorded all over Europe, eastward to Siberia.

Biology. *L. hippopus* feeds on plant fragments and other detritus. It occurs in both small and large streams, and has also been recorded from lakes. *L. hippopus* has a one-year life cycle. In areas with oceanic climate, emergence occurs during the first part of March, while in alpine areas and in northern Fennoscandia emergence may occur in June and July. It is able to form local populations that differ in the growth rate at the same water temperature (Lillehammer, 1987a).

42. ***Leuctra nigra*** (Olivier, 1811)
Figs 248, 252, 254, 257.

Leuctra nigra Olivier, 1811, Encycl. Methodique, 8: 186.
Leuctra acuminata Bengtsson, 1933, Lunds Univ. Årsskr., 2: 29.

Male. Body length 4.6-5.6 mm, wing length 5.0-6.1 mm. Body entirely black, wings smoky grey with darker veins. Terga 6, 7 and 8 divided, 6th and 8th segments each with a pair of well separated tergal processes (Fig. 248). Always long-winged.

Female. Body length 4.0-8.0 mm, wing length 5.7-10.9 mm. Body black except for the dorsal side of abdomen which is yellow to light brown. Subgenital plate consists of two main lateral lobes and a smaller central lobe (Fig. 252). Always long-winged.

Nymphs. First instar 0.75-0.80 mm. Fully grown nymphs 5-8 mm. Body light brown to yellow, with short legs and long antennae and cerci. Pronotum and the rest of the body fringed with long fine hairs (Fig. 254). Hind leg densely covered by long bristles (Fig. 257).

Variability. Less variation occurs in the male tergal processes of this species than in other *Leuctra* species. However, there may occur bilobed tergal processes also on segment 7. Little variation is seen in the genitalia of the females.

Distribution. *L. nigra* occurs over most of Fennoscandia. In Denmark it has been recorded only from one locality in Jutland and one locality on the island of Funen. *L. nigra* is also absent from the islands of Bornholm, Öland and Gotland. It occurs in most parts of Finland. The species is recorded in central and northern Europe, including parts of Great Britain.

Biology. The nymph feeds on plant fragments and other detritus. It occurs mainly in small streams, but may occasionally be taken in larger rivers. The species has a two-year life cycle, adults emerging in May-July (Thorup, 1973; Haaland, 1981).

		Germany	G. Britain	DENMARK											Sk.	Bl.
				SJ	EJ	WJ	NWJ	NEJ	F	LFM	SZ	NWZ	NEZ	B		
Arcynopteryx compacta (McL.)	1	●														
Diura bicaudata (L.)	2	●	●												●	
D. nanseni (Kempny)	3															
Isogenus nubecula Newm.	4	●														
Perlodes dispar (Ramb.)	5	●			●										●	●
P. microcephala (Pict.)	6	●	●		●	●		●								
Isoperla difformis (Klap.)	7	●		●	●	●	●	●							●	●
I. grammatica (Poda)	8	●	●	●	●	●	●	●	●		●		●		●	●
I. obscura (Zett.)	9	●													●	
Dinocras cephalotes (Curt.)	10	●	●		●	●									●	●
Isoptena serricornis (Pict.)	11	●			●	●										
Siphonoperla burmeisteri (Pict.)	12	●		●	●	●										
Xanthoperla apicalis (Newm.)	13	●														
Taeniopteryx nebulosa (L.)	14	●	●	●	●	●		●	●						●	●
Brachyptera braueri (Klap.)	15	●			●										●	●
B. risi (Mort.)	16	●	●	●	●			●	●		●			●	●	●
Rhabdiopteryx acuminata Klap.	17	●														
Amphinemura borealis (Mort.)	18	●													●	
A. palmeni Kop.	19															
A. standfussi (Ris)	20	●	●	●	●	●		●	●		●				●	●
A. sulcicollis (Stph.)	21	●	●		●	●		●							●	●
Nemoura arctica Esben-P.	22															
N. avicularis Mort.	23	●	●	●	●	●		●			●		●		●	●
N. cinerea (Retz.)	24	●	●	●	●	●	●	●	●	●	●	●	●	●	●	●
N. dubitans Mort.	25	●	●	●	●							●	●		●	
N. flexuosa Aub.	26	●		●	●	●		●	●		●	●	●		●	●
N. sahlbergi Mort.	27															
N. viki Lilleh.	28															
Nemurella pictetii Klap.	29	●	●	●	●	●	●	●	●		●		●		●	●
Protonemura hrabei Raušer	30	●			●			●								
P. intricata (Ris)	31	●														
P. meyeri (Pict.)	32	●	●	●	●	●		●							●	●
Capnia atra Mort.	33	●														
C. bifrons (Newm.)	34	●	●	●	●			●	●				●	●	●	
C. nigra (Pict.)	35	●													●	
C. pygmaea (Zett.)	36															
C. vidua Klap.	37	●	●													
Capnopsis schilleri (Rost.)	38	●													●	

SWEDEN

	Hall.	Sm.	Öl.	Gtl.	G. Sand.	Ög.	Vg.	Boh.	Dlsl.	Nrk.	Sdm.	Upl.	Vstm.	Vrm.	Dlr.	Gstr.	Hls.	Med.	Hrj.	Jmt.	Ång.	Vb.	Nb.	Ås. Lpm.	Ly. Lpm.	P. Lpm.	Lu. Lpm.	T. Lpm.
1																			●	●					●	●	●	●
2		●				●	●								●		●		●	●			●	●	●	●	●	●
3														●	●			●	●	●	●	●	●	●	●		●	●
4												●		●			●	●	●	●		●	●		●			
5	●	●				●	●	●	●	●			●		●	●	●		●	●								
6																												
7	●	●					●	●	●						●				●	●							●	
8	●	●				●	●	●	●	●		●	●	●	●	●	●	●	●	●	●	●	●	●	●		●	●
9	●	●					●					●		●	●	●	●	●	●	●		●	●	●	●		●	●
10	●	●					●								●				●	●					●		●	
11																											●	
12									●	●			●	●	●		●		●	●	●			●	●		●	●
13														●	●					●		●	●					
14	●	●				●		●	●			●	●	●	●			●	●	●	●	●			●		●	●
15	●	●					●																					
16	●	●						●		●		●	●	●	●				●	●					●		●	●
17																												
18	●	●				●	●	●	●	●	●	●			●		●	●	●	●		●	●	●	●		●	
19																												
20		●								●					●		●	●		●					●	●	●	●
21	●	●					●	●	●	●	●	●		●	●	●	●		●	●	●	●			●		●	●
22																							●				●	●
23		●					●			●					●	●			●	●						●	●	●
24	●	●	●	●		●	●	●	●	●	●	●	●	●	●	●	●	●	●	●	●	●	●	●	●	●	●	●
25															●													
26		●				●	●			●					●										●			
27																											●	●
28																												
29	●	●				●	●	●	●	●		●	●	●	●	●			●	●	●	●		●	●		●	●
30																												
31																												
32	●	●					●	●		●		●			●				●	●					●		●	
33		●				●	●								●				●	●	●		●		●	●	●	●
34							●																					
35																												
36															●				●	●	●		●		●		●	●
37																												●
38										●					●				●	●						●	●	

		NORWAY														
		Ø + AK	HE (s + n)	O (s + n)	B (ø + v)	VE	TE (y + i)	AA (y + i)	VA (y + i)	R (y + i)	HO (y + i)	SF (y + i)	MR (y + i)	ST (y + i)	NT (y + i)	Ns (y + i)
Arcynopteryx compacta (McL.)	1			●			●				●	●	●	●	●	
Diura bicaudata (L.)	2		●	●	●		●			●	●	●	●	●	●	●
D. nanseni (Kempny)	3	●	●	●	●	●	●	●	●	●	●	●	●	●	●	●
Isogenus nubecula Newm.	4		●	●	●											
Perlodes dispar (Ramb.)	5	●														
P. microcephala (Pict.)	6															
Isoperla difformis (Klap.)	7	●	●						●	●				●	●	●
I. grammatica (Poda)	8	●	●	●	●	●	●	●	●	●	●	●	●	●	●	●
I. obscura (Zett.)	9	●	●	●	●		●			●	●	●	●	●	●	●
Dinocras cephalotes (Curt.)	10		●	●			●			●	●			●	●	
Isoptena serricornis (Pict.)	11															
Siphonoperla burmeisteri (Pict.)	12	●	●	●	●	●	●	●	●	●	●	●	●	●	●	●
Xanthoperla apicalis (Newm.)	13		●											●		●
Taeniopteryx nebulosa (L.)	14	●	●	●	●	●	●	●	●	●	●	●	●	●	●	●
Brachyptera braueri (Klap.)	15															
B. risi (Mort.)	16	●	●	●	●	●	●	●	●	●	●	●	●	●	●	●
Rhabdiopteryx acuminata Klap.	17															
Amphinemura borealis (Mort.)	18	●	●	●	●	●	●	●	●	●	●	●	●	●	●	●
A. palmeni Kop.	19															
A. standfussi (Ris)	20	●	●	●	●		●			●	●	●	●	●	●	●
A. sulcicollis (Stph.)	21	●	●	●	●	●	●	●	●	●	●	●	●	●	●	●
Nemoura arctica Esben-P.	22															
N. avicularis Mort.	23	●	●	●	●	●	●	●	●		●			●	●	●
N. cinerea (Retz.)	24	●	●	●	●	●	●	●	●	●	●	●	●	●	●	●
N. dubitans Mort.	25															
N. flexuosa Aub.	26	●	●	●			●							●	●	●
N. sahlbergi Mort.	27															
N. viki Lilleh.	28															
Nemurella pictetii Klap.	29	●	●	●	●	●	●	●	●	●	●	●	●	●	●	●
Protonemura hrabei Raušer	30															
P. intricata (Ris)	31															
P. meyeri (Pict.)	32	●	●	●	●	●	●	●	●	●	●	●	●	●	●	●
Capnia atra Mort.	33	●	●	●	●		●		●	●	●	●	●	●	●	●
C. bifrons (Newm.)	34	●	●	●	●	●	●								●	●
C. nigra (Pict.)	35															
C. pygmaea (Zett.)	36		●	●			●		●	●	●	●	●	●	●	●
C. vidua Klap.	37															
Capnopsis schilleri (Rost.)	38	●	●	●	●	●	●							●	●	●

					FINLAND																				USSR		
	Nn (ø + v)	TR (y + i)	F (v + i)	F (n + ø)	Al	Ab	N	Ka	St	Ta	Sa	Oa	Tb	Sb	Kb	Om	Ok	Ob S	Ob N	Ks	LkW	LkE	Le	Li	Vib	Kr	Lr
1	●	●	●	●																			●	●			●
2	●	●	●	●				●		●	●		●	●	●		●			●	●	●	●	●	●	●	●
3	●	●	●	●			●		●	●				●			●	●	●	●	●	●	●	●			●
4							●	●										●	●	●	●	●		●	●	●	●
5								●																	●	●	
6																											
7			●	●			●			●				●						●			●	●		●	
8	●	●	●	●		●	●			●	●		●	●			●	●	●	●	●	●	●	●	●	●	●
9	●	●	●	●			●		●		●			●	●	●	●	●	●	●	●	●	●	●	●	●	●
10			●																								
11																			●		●	●		●			
12		●	●	●			●						●				●	●	●	●	●	●	●	●	●	●	●
13				●													●		●	●	●	●	●	●			●
14	●	●	●	●		●	●	●		●		●		●		●	●	●	●	●	●	●	●	●	●		●
15																											
16	●	●	●	●																				●			
17							●								●												
18		●	●	●		●	●		●	●	●		●	●		●	●	●	●	●	●	●	●	●	●	●	●
19				●																				●			●
20		●	●	●							●			●			●	●	●	●	●	●	●	●		●	●
21	●	●	●	●													●			●	●	●	●	●			
22		●	●	●														●	●	●	●	●	●	●			●
23	●	●	●	●		●	●			●			●	●			●	●	●	●	●	●	●	●		●	●
24	●	●	●	●	●	●	●	●	●	●	●	●	●	●	●	●	●	●	●	●	●	●	●	●	●	●	●
25							●		●																		
26	●	●	●	●		●		●		●							●						●	●		●	
27	●	●	●	●																			●	●			●
28			●	●																			●	●			
29	●	●	●	●		●	●	●		●	●			●	●		●	●	●	●	●	●	●	●	●	●	●
30																											
31				●											●					●				●		●	●
32	●	●	●	●																	●	●	●	●		●	●
33	●	●	●	●			●		●	●		●			●	●	●	●	●	●	●	●	●	●		●	●
34																											
35																											
36	●	●	●	●													●						●	●			●
37	●	●	●	●																			●	●			
38	●	●	●	●		●	●	●		●	●			●	●		●			●	●	●	●	●	●	●	●

		DENMARK														
		Germany	G. Britain	SJ	EJ	WJ	NWJ	NEJ	F	LFM	SZ	NWZ	NEZ	B	Sk.	Bl.
Leuctra digitata Kempny	39	●		●	●	●		●								
L. fusca (L.)	40	●	●	●	●	●		●	●		●			●	●	●
L. hippopus Kempny	41	●	●	●	●	●		●			●		●	●	●	●
L. nigra (Oliv.)	42	●	●	●	●	●		●	●						●	●

		NORWAY														
		Ø + AK	HE (s + n)	O (s + n)	B (ø + v)	VE	TE (y + i)	AA (y + i)	VA (y + i)	R (y + i)	HO (y + i)	SF (y + i)	MR (y + i)	ST (y + i)	NT (y + i)	Ns (y + i)
Leuctra digitata Kempny	39	●	●	●	●		●			●	●	●		●	●	●
L. fusca (L.)	40	●	●	●	●	●	●	●	●	●	●	●	●	●	●	●
L. hippopus Kempny	41	●	●	●	●	●	●	●	●	●	●	●	●	●	●	●
L. nigra (Oliv.)	42	●	●	●	●	●	●	●	●	●	●	●	●	●	●	●

SWEDEN

	Hall.	Sm.	Öl.	Gtl.	G. Sand.	Ög.	Vg.	Boh.	Dlsl.	Nrk.	Sdm.	Upl.	Vstm.	Vrm.	Dlr.	Gstr.	Hls.	Med.	Hrj.	Jmt.	Ång.	Vb.	Nb.	Ås. Lpm.	Ly. Lpm.	P. Lpm.	Lu. Lpm.	T. Lpm.
39														●	●				●	●	●				●		●	●
40	●	●				●	●	●	●			●	●	●	●	●	●		●	●	●	●			●		●	
41	●	●				●	●			●		●		●	●		●		●	●		●			●		●	●
42	●	●					●			●				●	●				●	●					●		●	●

					FINLAND																				USSR		
	Nn (ø + v)	TR (y + i)	F (v + i)	F (n + ø)	Al	Ab	N	Ka	St	Ta	Sa	Oa	Tb	Sb	Kb	Om	Ok	Ob S	Ob N	Ks	LkW	LkE	Le	Li	Vib	Kr	Lr
39	●	●	●	●		●	●	●		●	●			●	●					●	●	●	●	●	●	●	●
40	●	●	●	●		●	●	●	●	●			●	●	●		●	●	●	●	●	●	●	●	●	●	●
41	●	●	●	●										●	●		●	●		●	●		●	●	●	●	●
42	●	●	●	●		●	●			●					●		●	●	●	●	●	●	●	●		●	●

References

Anderson, N. H. & Cummins, K. W., 1979. Influence of diet on the life histories of aquatic insects. - J. Fish. Res. Bd Can., 36: 335-342.

Aubert, J., 1946. Les Plécoptères de la Suisse Romande. - Mitt. schweiz. ent. Ges., 20: 8-128.

- 1949. Plécoptères helvétiques. Notes morphologiques et systématiques. - Mitt. schweiz. ent. Ges., 22: 217-236.

- 1959. Plecoptera. - Insecta Helvetica, 1: 1-140.

Baekken, T., 1981. Growth patterns and food habits of *Baetis rhodani, Capnia pygmaea* and *Diura nanseni* in a west Norwegian river. - Holarct. Ecol., 4: 139-144.

Bagge, P., 1965. Observations on some mayfly and stonefly nymphs (Ephemeroptera and Plecoptera) in Utsjoki, Finnish Lappland. - Annls Ent. Fenn., 31: 102-108.

Bagge, P. & Salmela, V. M., 1978. The macrobenthos of the River Taurujoki and its tributaries (Central Finland). 1. Plecoptera, Ephemeroptera and Trichoptera. - Notul. ent., 58: 159-168.

Banks, N., 1903. New name for *Dictyopteryx* Pictet. - Ent. News, 14: 241.

- 1906. On the perlid genus *Chloroperla*. - Ent. News, 17: 174-175.

- 1914. Perlidae. *In:* New neuropteroid insects, native and exotic. - Proc. Acad. nat. Sci. Philad., 66: 608-611.

Benedetto, L. A., 1973a. Notes on North Swedish Plecoptera. - Ent. Tidskr., 94: 20-22.

- 1973b. Growth of stonefly nymphs in Swedish Lapland. - Ent. Tidskr., 94: 15-19.

Bengtsson, J., 1972. Vækst og livscyklus hos *Nemoura cinerea* (Retz.) (Plecoptera). - Flora Fauna, 78: 97-101.

- 1979. Klækningsmønster hos *Nemoura cinerea* (Retz.) (Plecoptera). - Flora Fauna, 85: 83-86.

Bengtsson, S., 1931. 12. Sjösländor - Plecoptera. *In:* Insektfaunan inom Abisko Nationalpark III. (by Y. Sjöstedt *et al.*). - K. svenska VetenskAkad. Skr. Naturskydd., 18: 1-72.

- 1933. Plecopterologische Studien. - Lunds Univ. Årsskr. (2), 29: 1-50.

Berg, R., 1963. Disjunksjoner i Norges fjellflora og de teorier som er fremsatt til forklaring av dem. Disjunctions in the Norwegian Alpino Flora and Theories Proposed for their Explanation. - Blyttia, 21: 133-177.

Billberg, G. J., 1820. Enumeratio Insectorum in Museo Billberg. Holmiae.

Brekke, R., 1941. The Norwegian Stoneflies, Plecoptera. - Norsk ent. Tidsskr., 6: 1-24.

Brinck, P., 1949. Studies on Swedish stoneflies (Plecoptera). - Opusc. ent., Suppl. 11: 1-250.

- 1952. Bäcksländor, Plecoptera. - Svensk Insektfauna, 15: 1-126.

- 1954. On the classification of the Plecopteran subfamily Perlodinae. - Opusc. ent., 19: 190-201.

- 1956. Reproductive system and mating in Plecoptera. - Opusc. ent., 21: 57-127.

- 1958. On some stoneflies recorded from Nowaya Zemlya. - Avh. norske VitenskAkad. Mat.-naturv. Kl. 2, 1958: 1-11.

- 1970. Plecoptera. Pp. 50-55 *in* Tuxen, S. L. (ed.): Taxonomist's glossary of genitalia in insects. Copenhagen.

Brinck, P. & Froehlich, C. G., 1960. On the stonefly fauna of western Lule Lappland, Swedish Lappland. - K. fysiogr. Sällsk. Lund Förh., 30 (1): 1-19.

Brinck, P. & Wingstrand, K. G., 1949. The mountain fauna of the Virihaure area in Swedish Lappland. I. General account. - Acta Univ. Lund. N. F., Avd. 2, 45 (2): 1-70.

- 1951. The mountain fauna of the Virihaure area in Swedish Lappland. II. Special account. - Acta Univ. Lund. N. F., Avd. 2, 45 (2): 1-173.

Brittain, J. E., 1973. The biology and life cycle of *Nemoura avicularis* Morton (Plecoptera). - Freshwat. Biol., 3: 199-210.

- 1974. Studies on the lentic Ephemeroptera and Plecoptera of southern Norway. - Norsk ent. Tidsskr., 21: 135-153.
- 1977. The effect of temperature on the egg incubation period of *Taeniopteryx nebulosa* (Plecoptera). - Oikos, 29: 302-305.
- 1978. Semivoltinism in mountain populations of *Nemurella pictetii* (Plecoptera). - Oikos, 30: 1-6.
- 1983a. The influence of temperature on nymphal growth rates in mountain stoneflies (Plecoptera). - Ecology, 64: 440-446.
- 1983b. The first record of the nymph of *Xanthoperla apicalis* (Newman) (Plecoptera, Chloroperlidae) from Scandinavia, with a key to the mature nymphs of the Scandinavian Chloroperlidae. - Fauna norv., Ser. B., 30: 52-53.

Brittain, J. E., Lillehammer, A. & Saltveit, S. J., 1984. The effect of temperature on intraspecific variation in egg biology and nymphal size in the stonefly *Capnia atra* (Plecoptera). - J. Anim. Ecol., 53: 161-169.
- 1986. Intraspecific variation in the nymphal growth rate of the stonefly, *Capnia atra* (Plecoptera). - J. Anim. Ecol., 55: 1001-06.

Brock, W., 1986. Vergleichenden Untersuchungen zur Entwicklung der Steinfliegen (Insecta, Plecoptera). Dissertation zur Erlang des Doctorgrades des Fachbereiches Biologie der Universität Hamburg (1986), 149 pp.

Bronsky, A. K., 1981. Evolution of the wing apparatus in stoneflies (Plecoptera). III. Wing deformation on *Isogenus nubecula* Newman during flight. - Ent. Rev., 60: 25-36.

Claassen, P. W., 1936. New names for stoneflies (Plecoptera). - Ann. ent. Soc. Am., 29: 622-623.

Consiglio, C., 1967. Lista dei Plecotteri della regione Italiana. - Fragm. ent., 51: 1-66.

Cummins, K. W. & Klug, M. J., 1979. Feeding ecology of stream invertebrates. - Ann. Rev. Ecol. Syst., 10: 147-172.

Daan, S. & Gustavsson, K., 1973. Midsummer night emergence of stoneflies (Plecoptera) in a Lappland mountain lake. - Aquilo, Ser. Zool., 14: 29-33.

De Geer, C., 1778. Mémoires pour servir à l'histoire des Insectes. Vol. 7, Stockholm.

Dodds, G. S. & Hirshaw, F. L., 1925. Ecological studies on aquatic insects. IV. Altitudinal range and zonation of mayfly, stoneflies and caddisflies in the Colorado Rockies. - Ecology, 6: 380-390.

Eglishaw, H. J., 1964. The distributional relationship between the bottom fauna and plant detritus in streams. - J. Anim. Ecol., 34: 233-251.

Elliott, J. M., 1984. Hatching time and growth of *Nemurella pictetii* (Plecoptera: Nemouridae) in the laboratory and a Lake District stream. - Freshwat. Biol., 14: 491-499.
- 1986. The effect of temperature on the egg incubation period of *Capnia bifrons* (Plecoptera: Capniidae) from Windermere (English Lake District). - Holarct. Ecol., 9: 113-116.

Elvang, D. & Madsen, B. L., 1973. Biologiske undersøgelser over *Taeniopteryx nebulosa* (L.) (Plecoptera) med bemærkninger over vækst. - Ent. Meddr, 44: 49-59.

Enderlein, G., 1905. Die Plecopteren Feuerlands. - Zool. Anz., 28: 809-815.
- 1909. Klassifikation der Plecopteren sowie Diagnosen neuer Gattungen und Arten. - Zool. Anz., 34: 385-419.

Esben-Petersen, P., 1910a. Plecoptera (Perlidae). Bidrag til en fortegnelse over arktisk Norges Neuropter-fauna. - Tromsø Mus. Årsh., 31/32: 82-86.
- 1910b. Guldsmede, Døgnfluer, Slørvinger og Copeognather. - Danmarks Fauna, 8.

Fuller, R. L. & Stewart, K. W., 1977. The food habits of stoneflies (Plecoptera) in the Upper Gunnison River, Colorado. - Envir. Ecol., 6: 293-302.
- 1979. Stonefly (Plecoptera). Food habits and prey preference in the Dolores River, Colorado. - Am. Midl. Nat., 101: 170-181.

Frost, S., Huni, A. & Kershaw, W. E., 1971. Evaluation of a kicking technique for sampling stream bottom fauna. - Can. J. Zool., 49: 167-183.
Haaland, Ø., 1981. Livssyklusstudier av steinfluene *Capnopsis schilleri* (Capniidae) og *Leuctra nigra* (Leuctridae) (Plecoptera). Unpublished thesis, University of Oslo. 55 pp.
Harper, P. P., 1973. Life histories of Nemouridae and Leuctridae in southern Ontario (Plecoptera). - Hydrobiologia, 41: 309-356.
Holdsworth, R. P., 1941. The life history of *Pteronarcys proteus* Newman. - Ann. ent. Soc. Am., 34: 495-502.
Huusko, H. & Kuusela, K., 1985. Plecoptera caught by an application of tree-eclector in a forest near a small river, eastern Finland. - Notul. ent., 65: 93-96.
Hynes, H. B. N., 1941. The taxonomy and ecology of the nymphs of British Plecoptera with notes on the adults and eggs. - Trans. R. ent. Soc. Lond. (A), 43: 40-48.
- 1953. The Plecoptera of some small streams near Silkeborg, Jutland. - Ent. Meddr, 26: 489-494.
- 1963. The gill-less Nemourid nymphs of Britain (Plecoptera). - Proc. R. ent. Soc. Lond. (A), 38: 70-76.
- 1970. The ecology of running waters. 555 pp., Liverpool.
- 1977. A key to the adults and nymphs of British stoneflies (Plecoptera) with notes on their ecology and distribution. - F. B. A. scient. Publ., 17: 1-90.
Ihomononow, P., 1973. Distribution saisonaire des Plécoptères (Insecta) dans eaux de la Montague char. - Godisen Zb. Skopje Biol., 25: 11-39.
Illies, J., 1953. Beitrag zur Verbreitungsgeschichte der europäischen Plecopteren. - Arch. Hydrobiol., 48: 35-74.
- 1955. Die Bedeutung der Plecopteren für die Verbreitungsgeschichte der Süsswasserorganismen. - Verh. int. Verein theor. angew. Limnol., 12: 643-653.
- 1962. Das abdominale Zentralnervensystem der Insekten und seine Bedeutung für Phylogenie und Systematik der Plecopteren. - TagBer. Versamml. dt. Ent., 45: 139-152.
- 1965. Phylogeny and zoogeography of the Plecoptera. - A. Rev. Ent., 10: 117-140.
- 1966. Katalog der rezenten Plecoptera. - Tierreich, 82: 1-632.
- 1967. Plecoptera. - Limnofauna europapaea. 474 pp. Stuttgart.
- 1978. Plecoptera. - Limnofauna europaea. 522 pp. Stuttgart.
Iversen, T. M., 1978. Life cycle and growth of three species of Plecoptera in Danish springs. - Ent. Meddr, 46: 57-62.
Jensen, C. F., 1951. Plecoptera (Slørvinger). En faunistisk-biologisk undersøgelse af Skern Å. I. - Flora Fauna, 57: 17-40.
Jensen, F., Jensen, C. F. & Munk, T., 1986. Nye fund fra Danmark af slørvingen *Protonemura hrabei* Raušer, 1956, samt nogle biologiske iagttagelser. - Flora Fauna, 92: 13-16.
Johnsson, C. G., 1966. A functional system of adaptive dispersal by flight. - A. Rev. Ent., 11: 233-260.
Kaiser, E. W., 1972. Status over de danske Plecoptera (Slørvinger), pp. 98-100, *in:* Status over den danske dyreverden. - Symp. ved Københavns Universitet, 1971.
Kempny, P., 1898. Zur Kenntnis der Plecopteren. I. Ueber *Nemura* Latr. - Verh. zool.-bot. Ges. Wien, 48: 37-68.
- 1899. Zur Kenntnis der Plecopteren. II: Neue und ungenügend bekannte *Leuctra*-Arten. - Verh. zool.-bot. Ges. Wien, 49: 9-15, 269-278.
- 1900. Ueber die Perliden-Fauna Norwegens. - Verh. zool.-bot. Ges. Wien, 50: 85-99.
Khoo, S. G., 1968a. Experimental studies on diapause in stoneflies. I: Nymphs of *Capnia bifrons* (Newman). - Proc. R. ent. Soc. Lond. (A), 43: 40-48.
- 1968b. Experimental studies on diapause in stoneflies. II. Eggs of *Diura bicaudata* (L.). - Proc.

R. ent. Soc. Lond. (A), 43: 49-56.
Kimmins, D. E., 1943. *Rhabdiopteryx anglica,* and new British species of Plecoptera. - Proc. R. ent. Soc. Lond. (B), 12: 42-44.
- 1950. Plecoptera. - Handbk Ident. Br. Insects 1 (6): 18 pp. London.
Klapálek, F., 1900. Plekopterologické studie. - Rozpr. české Akad. Cis. Fr. Jos., II, 9: 1-34.
- 1902. Zur Kenntnis der Neuropteroiden von Ungarn, Bosnien und Herzegovina. - Természetr. Füz., 25: 161-180.
- 1904a. O vnějších plodidlech ♂ *Arcynopteryx dovrensis* Mort. - Čas. české Spol. ent., 1: 104-106.
- 1904b. Ueber die europaeischen Arten der Fam. Dictyopterygidae. - Bull. int. Acad. Sci. Bohème, 9: 6-15.
- 1904c. Zpráva o výsledcích cesty do Transsylvanskych Alp a Vysokých Tater. - Rozpr. české Akad. Cís Fr. Jos., II, 13: 722-724.
- 1905a. Přispěvek K rodu *Rhabdiopteryx* Klp. - Čas. české Spol. ent., 2: 10-14.
- 1905b. Conspectus Plecopterorum Bohemiae. - Čas. české Spol. ent., 2: 27-32.
- 1907a. *Taeniopteryx dusmeti* Navás a *T. ornata* Navás. - Čas. české Spol. ent., 4: 23-24.
- 1907b. Evropské druhy rodu *Perla* Geoffr. - Rozpr. české Akad. Cís Fr. Jos., II, 16: 1-25.
- 1909a. Plecoptera, Steinfliegen. *In:* Brauer, Die Süsswasserfauna Deutschlands, 8: 33-95.
- 1909b. *Capnia conica* n.sp. - Čas. české Spol. ent., 6: 101-102.
Knight, A. W. & Gaufin, A. R., 1966. Altitudinal distribution of stoneflies (Plecoptera) in a Rocky Mountain drainage system. - J. Kansas ent. Soc., 39: 668-675.
Koponen, J. S. W., 1915. Suomen koskikorennoisista. - Meddn Soc. Fauna Flora fenn., 41: 24-26.
- 1917. Plecopterologische Studien. I. Die Plecopteren-Arten Finnlands. - Acta Soc. Fauna Flora fenn., 44: 1-18.
- 1949. Neue oder wenig bekannte Plecoptera. - Annls. ent. fenn., 15: 1-21.
Kuusela, K., 1984. Emergence of Plecoptera in two lotic habitats in the Oularika National Park, north east of Finland. - Annls. Limnol., 20: 63-68.
Kuusela, K. & Pulkkinen, H., 1978. A simple trap for collecting newly emerged stoneflies (Plecoptera). - Oikos, 81: 323-325.
Kühtreiber, J., 1934. Die Plecopterfauna Nordtirols. - Ber. naturw.-med. Ver. Innsbruck, 43/44: 1-219.
Latreille, P. A., 1796. Précis des charactères géneriques des Insectes, disposés dans un ordre naturel. xiv + 201 pp. Paris.
Lillehammer, A., 1972a. Notes on the stonefly *Capnia vidua* Klapálek from Fennoscandia. - Norsk ent. Tidsskr., 19: 153-156.
- 1972b. Notes on the stonefly *Nemoura sahlbergi* Morton, with a description of the nymph. - Norsk ent. Tidsskr., 19: 157-159.
- 1972c. A new species of the genus *Nemoura* (Plecoptera) from Finnmark, North Norway. - Norsk. ent. Tidsskr., 19: 161-163.
- 1973. The nymph of *Capnopsis schilleri* (Rostock, 1892). Notes on its morphology and emergence. - Norsk ent. Tidsskr., 20: 267-268.
- 1974a. Norwegian stoneflies I. Analyses of the variations in morphological and structural characters used in taxonomy. - Norsk ent. Tidsskr., 21: 59-107.
- 1974b. Norwegian stoneflies II. Distribution and relationship to the environment. - Norsk ent. Tidsskr., 21: 159-250.
- 1975a. Norwegian stoneflies III. Field studies on ecological factors influencing distribution. - Norw. J. Ent., 22: 71-80.
- 1975b. Norwegian stoneflies IV. Laboratory studies on ecological factors influencing distribution. - Norw. J. Ent., 22: 99-108.

- 1976. Norwegian stoneflies V. Variations in morphological characters compared to differences in ecological factors. - Norw. J. Ent., 22: 161-172.
- 1978. The Plecoptera of Øvre Heimdalen. - Holarct. Ecol., 1: 232-238.
- 1984. Distribution, seasonal abundance and emergence of stoneflies (Plecoptera) in the Øvre Heimdal area of the Norwegian Jotunheim Mountains. - Fauna norv., Ser. B., 31: 1-7.
- 1985a. Zoogeographical studies on Fennoscandian stoneflies (Plecoptera). - J. Biogeogr., 12: 209-221.
- 1985b. Temperature influence on egg incubation period and nymphal growth on the stoneflies *Leuctra digitata* and *L. fusca* (Plecoptera: Leuctridae). - Entomologia gen., 11: 59-67.
- 1985c. The coexistence of stoneflies in a mountain lake outlet biotop. - Aquat. Ins., 7: 173-187.
- 1985d. Studies of shortwingedness in stoneflies (Plecoptera). - Fauna norv., Ser. B., 32: 58-61.
- 1986a. Taxonomic differences between populations of *Leuctra hippopus* Kempny (Plecoptera) in Norway. - Fauna norv., Ser. B., 33: 27-32.
- 1986b. The effect of temperature on egg incubation period and nymphal growth of two *Nemoura* species (Plecoptera) from subarctic Fennoscandia. - Aquat. Ins., 8: 223-235.
- 1986c. Studies on *Capnia vidua* Klapálek (Capniidae, Plecoptera) populations in Iceland. - Fauna norv., Ser. B., 33: 93-97.
- 1987a. Intraspecific variation in the biology of eggs and nymphs of Norwegian populations of *Leuctra hippopus* (Plecoptera). - J. nat. Hist., 21: 29-41.
- 1987b. Diapause and quiescence in eggs of Systellognatha stonefly species (Plecoptera). - Annls Limnol. 23 (3).
- 1987c. Egg development of the stoneflies *Siphonoperla burmeisteri* (Chloroperlidae) and *Dinocras cephalotes* (Perlidae). - Freshwat. Biol., 17: 35-39.
- 1987d. Taxonomy of the Fennoscandian *Nemoura* nymphs (Nemouridae, Plecoptera), with a key to species. - Ent. scand., 17: 511-19.

Lillehammer, A. & Brittain, J. E., 1978. The invertebrate fauna of the streams in Øvre Heimdalen. - Holarct. Ecol., 1: 271-276.

Lillehammer, A., Brittain, J. E. & Saltveit, S. J., 1988. Egg development, nymphal growth and life cycle strategies in Plecoptera. - Holarct. Ecol., 11.

Lillehammer, A. & Økland, B., 1987. Taxonomy of stonefly eggs of the genus *Isoperla* (Plecoptera, Perlodidae). - Fauna norv., Ser. B., 34: 121-124.

Lindroth, C., 1949. The theory of glacial refugia in Scandinavia. Comments on present opinions. - Notul. ent., 49: 178-192.

Linnaeus, C., 1758. Systema Naturae. Ed. 10. Holmiae.

Madsen, B. L., 1968. The distribution of nymphs of *Brachyptera risi* Mort. and *Nemoura flexuosa* Aub. (Plecoptera) in relation to oxygen. - Oikos, 19: 304-310.
- 1969. Reaction of *Brachyptera risi* (Morton) (Plecoptera) nymphs to water currents. - Oikos, 20: 95-100.
- 1974. A note on the food of *Amphinemura sulcicollis* (Plecoptera). - Hydrobiologia, 45: 169-175.

Malmquist, B. & Sjøstrøm, P., 1980. Prey size and pattern of feeding in *Dinocras cephalotes* (Plecoptera). - Oikos, 35: 311-316.

McLachlan, R., 1872. Pelides. *In:* Non-Odonates (Second part. Matériaux pour une fauna névroptérologique de l'Asia septentrionale), by DE SÉLYS-LONGCHAMPS and MCLACHLAN. - Annls Soc. ent. Belg., 15: 51-55.

Meinander, M., 1965. List of the Plecoptera of Eastern Fennoscandia. - Fauna Fennica, 19: 1-38.
- 1972. The invertebrate fauna of Kilpisjärvi area, Finnish Lappland 4. Plecoptera. - Acta Soc. Fauna Flora fenn., 80: 45-61.
- 1975. Notes on Plecoptera from Eastern Fennoscandia. - Notul. ent., 55: 129-130.

- 1980. Suomen koskikorennot - Finnlands bäckslädor (Plecoptera). - Notul. ent., 60: 7-10.
- 1984. Plecoptera of Inari Lapland. - Kevo Notes, 7: 39-40.
Morton, K. J., 1894. XXIII. Palaearctic Nemourae. - Trans. ent. Soc. Lond., 1894: 557-574.
- 1896. II. New and little-known Palaearctic Perlidae. - Trans. ent. Soc. Lond., 1896: 55-63.
- 1901. Perlidae taken in Norway in June and July, 1900, with remarks on certain arctic forms. - Entomologist's mon. Mag., 37: 146-148.
- 1930. Plecoptera collected in Corsica by Mr. Martin E. Mosely. - Entomologist's mon. Mag., 66: 75-81.
Müller, K., 1973. Life cycles of stream insects. - Aquilo, Ser. Zool., 14: 105-112.
- 1978. Vingeutveckling hos *Capnia atra* Morton i Abisko-området (Plecoptera). - Ent. Tidskr., 99: 111-113.
Müller, K. & Mendl, H., 1980. On the biology of the stonefly species *Leuctra digitata* in a northern Swedish coastal stream and its adjacent coastal areas (Plecoptera: Leuctridae). - Entomologica gen., 6: 97-105.
Navás, R. P. L., 1903. (*Taeniopteryx*). *In:* Algunos Neuropteros de España nuevos. - Boln Soc. aragon. Cienc. nat., 2: 102-106.
- 1917a. Familia Pérlidos. *In:* Neurópteros nuevos de España. - Revta R. Acad. Cienc. Madr., 15: 741-742.
- 1917b. Plecópteros. *In:* Neurópteros nuevos o poco conocidos. - Mems R. Acad. Cienc. Artes Barcelona, 13: 5-8 (157-160).
- 1918. Plecópteros. *In:* Neurópteros nuevos o poco conocidos. - Mems R. Acad. Cienc. Artes Barcelona, 14: 5-13 (341-349).
- 1927. Plecópteros. *In:* Insectos nuevos de la Peninsula Ibérica. - Boln Soc. ent. Esp., 10: 82-83.
- 1930. Entomologische Ergebnisse der schwedischen Kamtchatka-Expedition 1920-1922. Plecoptera. - Ark. Zool., 21(A7): 1-8.
- 1932. Plecópteros. *In:* Insectos de Francia interesantes. - Boln Soc. ent. Esp., 15: 93-96.
Nebeker, A. V., 1971. Effect of temperature at different altitudes on the emergence of aquatic insects from a single stream. - J. Kans. ent. Soc., 44: 26-35.
Nelson, C. H., 1984. Numerical cladistic analysis of phylogenetic relationship in Plecoptera. - Ann. ent. Soc. Am., 77: 466-473.
Nelson, C. H. & Garth, R. E., 1984. Oxygen consumption of several species of Plecoptera. - J. Tenn. Acad. Sci., 59: 27-28.
Newman, E., 1833. *Isogenus. In:* Entomological notes. - Ent. Mag., London, 1: 415.
- 1836. Entomological notes. - Ent. Mag., London, 3: 499-501.
- 1839. On the synonymy of Perlites, together with brief characters of the old, and of a few new species. - Ann. Mag. nat. Hist., (2), 3: 32-37, 84-90.
Newport, G., 1848-49. Postscript to Mr. Newport's paper on *Pteronarchys regalis*. - Proc. linn. Soc. Lond., 1: 387-389.
Nordal, I., 1985a. Overvintringsteori og evolusjonshastighet. - Blyttia, 43: 33-42.
- 1985b. Overvintringsteori og det vestarctiske element i skandinavisk flora. - Blyttia, 43: 185-193.
Olivier, G. A., 1811. Némoure. *In:* Encyclopédie méthodique. Dictionnaire des Insectes. Vol. 8. Paris.
Petterson, C., 1983. Aims and methods in biogeography. Pp. 1-28, *in:* Sims, R. W., Price, J. H. & Whaley, P. E. S.: Evolution, time and space: The emergence of the biosphere. London.
Pianka, E. R ., 1981. Competition and niche theory. Pp. 167-196, *in:* May, R. M. (ed.) - Theoretical Ecology. Principles and Applications. Oxford.
Pictet, A. E., 1865. Névroptères d'Espagne. Genève.
Pictet, F. J., 1833. Mémoire sur les métamorphoses des Perles. - Annls Sci. nat., 28: 44-65.

- 1841. Histoire naturelle générale et particulière des insectes Névroptères. Famille des Perlides. - 1. Partie: 1-423. Genève.
Poda, N., 1761. Insecta Musei Graecensis. Widmanstad.
Rambur, J. P., 1842. Tribu des Perlides. *In:* Histoire naturelle des Insectes. Névroptères: 449-462. Paris.
Raušer, J., 1956. Zur Kenntnis der tschechoslowakischen *Protonemura*-Larven. - Acta Akad. Sci. Čsl., Brno, 28: 449-496.
- 1962. Zur Verbreitungsgeschichte einer Insektdauerngruppe (Plecoptera) in Europa. - Pr. brn. Zakl. čsl. Akad. Věd, 34: 281-283.
- 1968. Plecoptera. Ergebnisse der zoologischer Forschung von Dr. Z. Kaszab in der Mongolei. - Ent. Abh. Mus. Tierk. Dresden, 34: 329-398.
- 1971. A contribution to the question of the distribution and evolution at plecopterological communities in Europe. - Acta faun. ent. Mus. nat. Pragae, 14: 33-63.
Rekstad, O., 1979. Vekst- og livssyklusstudier av tre steinfluer (fam. Nemouridae) fra Sørkedalen. - Unpubl. thesis, University of Oslo. 46 pp.
Retzius, A. I., 1783. Caroli de Geer genera et species insectorum: 1-220. Lipsiae.
Ris, F., 1902. Die schweizerischen Arten der Perliden-Gattung *Nemoura*. - Mitt. schweiz. ent. Ges., 10: 378-405.
Rognes, K., 1986. The West-Arctic distribution pattern. - Blyttia, 44: 76-81.
Rostock, M., 1892. *Capnodes schilleri,* eine neue deutsche Perlide. - Berl. ent. Z., 37: 1-5.
Ruprecht, R., 1969. Zur Artsspezifität der Trommelsignale der Plecopteren (Insecta). - Oikos, 20: 26-33.
- 1972. Dialektbildung bei den Trommelsignalen von *Diura* (Plecoptera). - Oikos, 23: 410-412.
- 1976. Struktur und Function der Bauchblase und der Hammers von Plecoptera. - Zool. Jb. (Anat.), 95: 9-80.
- 1982. Drumming signals of Danish Plecoptera. - Aquat. Ins., 4: 93-103.
Ruprecht, R. & Gnatzy, W., 1974. Die Feinstrukture der Sinneshaare auf der Bauchblasen von *Leuctra hippopus* und *Nemoura cinerea* (Plecoptera). - Cytobiologie, 9: 422-?
Saltveit, S. J., 1977. Felt- og laboratoriestudier på steinfluer (Plecoptera) i Sørkedalselven med spesiell vekt på slekten *Amphinemura*. - Unpubl. thesis, Univ. Oslo.
- 1978. The small nymphs of *Diura nanseni* (Kempny) (Plecoptera). - Ent. scand., 9: 297-298.
Saltveit, S. J. & Lillehammer, A., 1984. Studies on egg development in the Fennoscandian *Isoperla* species. - Annls Limnol., 20: 91-94.
Saltveit, S. J. & Brittain, J. E., 1986. Shortwingedness in the stonefly *Diura nanseni* (Kempny) (Plecoptera: Perlodidae). - Ent. scand., 17: 153-156.
Schoenemund, E., 1912. Zur Biologie und Morphologie einiger *Perla*-Arten. - Zool. Jb. (Anat.), 34: 1-56.
Siegfrid, C. A. & Knight, A. W., 1976. Prey selection by a setipalpian stonefly nymph *Acroneuria californica* Banks (Plecoptera, Perlodidae). - Ecology, 57: 603-608.
Septon, D. H. & Hynes, H. B. N., 1982. The numbers of nymphal instars of several Australian Plecoptera. - Aquat. Ins., 4: 153-166.
Sheldon, A. L., 1980a. Resource division by Perlid stoneflies (Plecoptera) in a lake outlet ecosystem. - Hydrobiologia, 71: 155-161.
- 1980b. Coexistence of Perlid stoneflies (Plecoptera): Prediction from multivariable morphometrics. - Hydrobiologia, 71: 99-105.
Smith, L. M., 1917. Studies of North American Plecoptera (Pteronarcinae and Perlodini). - Trans. Am. ent. Soc., 43: 433-489.
Stark, B. P. & Szczytko, S. W., 1982. Egg morphology and polygeny in Pteronarcyidae (Plecoptera). - Ann. ent. Soc. Am., 75: 519-529.

Stark, B. P., Gonzales, Del Tanago, M. & Szczytko, S. W., 1986. Systematic studies on Western Palearctic Perlodini (Plecoptera: Perlodidae). - Aquat. Ins., 8: 91-98.
Stephens, J. F., 1836. Illustrations of British Entomology. Mandibulata. Vol. 6. London.
Stewart, B. P., Szczytko, S. W. & Stark, B. P., 1983. The language of Stoneflies. - Bioscience, 33: 117-118.
Svensson, P. O., 1966. Growth of nymphs of stream living stoneflies (Plecoptera) in northern Sweden. - Oikos, 17: 197-206.
Tauber, C. A. & Tauber, M. J., 1981. Insect seasonal cycles: Genetic and Evolution. - A. Rev. Ecol. Syst., 12: 281-308.
Thorup, J., 1967. *Protonemura hrabei* Raušer, ny for Danmark. - Flora Fauna, 83: 7-10.
- 1973. Interpretation of growth-curves for animals from Danish springs. - Hydrobiologia, 22: 55-84.
Thomas, E., 1966. Orientierung der Imagines von *Capnia atra* Morton (Plecoptera). - Oikos, 17: 278-280.
Tobias, D., 1973. Köcherfliegen und Steinfliegen einiger Gewässer in Sör-Varanger (Nord-Norwegen) (Trichoptera, Plecoptera). II. *Amphinemura norvegica* n.sp. (Nemouridae). - Senckenberg. biol., 54: 339-342.
- 1974. Köcherfliegen und Steinfliegen einiger Gewässer in Sör-Varanger (Nord-Norwegen) (Trichoptera, Plecoptera). III. Liste der gefundenen Steinfliegen-Arten. - Senckenberg. biol., 55: 165-168.
Ulfstrand, S., 1968. Life cycles of benthic insects in Lapland streams (Ephemeroptera, Plecoptera, Trichoptera, Diptera, Simuliidae). - Oikos, 19: 167-190.
- 1975. Diversity and some other parameters of Ephemeroptera and Plecoptera communities in subarctic running waters. - Arch. Hydrobiol., 76: 499-520.
Ulfstrand, S., Svensson, B., Enckell, P. H., Hagerman, L. & Otto, C., 1971. Benthic insect communities of streams in Stora Sjöfallet National Park, Swedish Lapland. - Ent. scand., 2: 309-336.
Vasiliu, G. D. & Costea, E., 1942. Systematische Überprüfung der Steinfliegen (Plecoptera) Rumäniens und deren geographischen Ausdehnungsfläche. - Anal. Inst. Cerc. pisc. Rom., 1: 191-204.
Walker, F., 1852. Perlides. *In:* Catalogue of the specimens of neuropterous insects in the collection of the British Museum. Part I. (Pleryganides-Perlides): 136-192. London.
Ward, J. V. & Stanford, J. A., 1982. Thermal responses in the evolutionary ecology of aquatic insects. - A. Rev. Ent., 27: 97-117.
Winterbourn, J. J., 1974. The life histories, trophic relations and production of *Stenoperla prasina* (Plecoptera) and *Deleatidium* sp. (Ephemeroptera) in New Zealand rivers. - Freshwat. Biol., 4: 507-524.
Wu, C. F., 1923. Morphology, anatomy and ethology of *Nemoura*. - Bull. Lloyd Libr., Ent., 3: 1-46.
Zhiltsova, L. A., 1964. Plecoptera. pp. 177-200 in Bei-Bienko (ed.) Opred. Nosekom. ewrope - tschast. SSSR, Moscow, Leningrad, 1.
Zwick, P., 1967. Revision der Gattung *Chloroperla* Newman (Plecoptera). - Mitt. schweiz. ent. Ges., 40: 1-26.
- 1972. Die Plecopteren Pictet's und Burmeister's mit Angaben über weitere Arten (Insecta). - Rev. suisse Zool., 78: 1123-1194.
- 1973. Insecta: Plecoptera. Phylogenetisches System und Katalog. Das Tierreich, 94: I-XXXII + 1-465. Berlin.
- 1974. Das phylogenetische System der Plecoptera. - Entomologica germ., 1: 50-57.

- 1980. Diapause development of *Protonemura intricata* (Plecoptera - Nemouridae). - Verh. int. Verein theor. angew. Limnol., 21: 1607-1611.
- 1981. Das Mittelmeergebiet als Glaziales Refugium für Plecoptera. - Acta ent. jugosl., 17: 107-111.
- 1984. Geographische Rassen und Verbreitungsgeschichte von *Capnopsis schilleri* (Plecoptera, Capniidae). - Dt. ent. Z., 31: 1-7.

Author's address:
Zoologisk Museum
Sars gt. 1, Oslo 5
Norway

Index

Synonyms are given in italics. The number in bold refers to the main treatment of the taxon.

Printed in the United States
By Bookmasters